WORLDS OF FOOD

Worlds of Food

Place, Power, and Provenance in the Food Chain

Kevin Morgan, Terry Marsden, and
Jonathan Murdoch

OXFORD
UNIVERSITY PRESS

OXFORD
UNIVERSITY PRESS

Great Clarendon Street, Oxford OX2 6DP

Oxford University Press is a department of the University of Oxford.
It furthers the University's objective of excellence in research, scholarship,
and education by publishing worldwide in

Oxford New York

Auckland Cape Town Dar es Salaam Hong Kong Karachi
Kuala Lumpur Madrid Melbourne Mexico City Nairobi
New Delhi Shanghai Taipei Toronto

With offices in

Argentina Austria Brazil Chile Czech Republic France Greece
Guatemala Hungary Italy Japan Poland Portugal Singapore
South Korea Switzerland Thailand Turkey Ukraine Vietnam

Oxford is a registered trade mark of Oxford University Press
in the UK and in certain other countries

Published in the United States
by Oxford University Press Inc., New York

© Kevin Morgan, Terry Marsden, and Jonathan Murdoch 2006

The moral rights of the authors have been asserted

Database right Oxford University Press (maker)

First published 2006

British Library Cataloguing in Publication Data

Data available

Library of Congress Cataloging-in-Publication Data

Morgan, Kevin.
Worlds of food : place, power, and provenance in the food chain /
Kevin Morgan, Terry Marsden, and Jonathan Murdoch.
p. cm.—(Oxford geographical and environmental studies)
Includes bibliographical references.
ISBN 0–19–927158–5 (alk. paper)
1. Food supply. 2. Agricultural industries. 3. Food industry and trade. 4. Sustainable agriculture.
I. Marsden, Terry. II. Murdoch, Jonathan. III. Title. IV. Series.
HD9000.5.M675 2006
338.1—dc22
2005023275

Typeset by SPI Publishers Services
Printed in Great Britain
on acid-free paper by
Biddles Ltd., King's Lynn

ISBN 978–0–19–927158–0

EDITORS' PREFACE

Geography and environmental studies are two closely related and burgeoning fields of academic enquiry. Both have grown rapidly over the past few decades. At once catholic in its approach and yet strongly committed to a comprehensive understanding of the world, geography has focused upon the interaction between global and local phenomena. Environmental studies, on the other hand, have shared with the discipline of geography an engagement with different disciplines, addressing wide-ranging and significant environmental issues in the scientific community and the policy community. From the analysis of climate change and physical environmental processes to the cultural dislocations of postmodernism in human geography, these two fields of enquiry have been at the forefront of attempts to comprehend transformations taking place in the world, manifesting themselves as a variety of separate but interrelated spatial scales.

The *Oxford Geographical and Environmental Studies* series aims to reflect this diversity and engagement. Our goal is to publish the best original research in the two related fields, and, in doing so, to demonstrate the significance of geographical and environmental perspectives for understanding the contemporary world. As a consequence, our scope is deliberately international and ranges widely in terms of topics, approaches, and methodologies. Authors are welcome from all corners of the globe. We hope the series will help to redefine the frontiers of knowledge and build bridges within the fields of geography and environmental studies. We hope also that it will cement links with issues and approaches that have originated outside the strict confines of these disciplines. In doing so, our publications contribute to the frontiers of research and knowledge while representing the fruits of particular and diverse scholarly traditions.

Gordon L. Clark
Andrew Goudie
Ceri Peach

ACKNOWLEDGEMENTS

Some books are more of a collective endeavour than others, and this is emphatically one of them. From start to finish we have received a wide array of support—intellectual, practical, and emotional. In the United States we must thank Bill Friedland, Melanie DuPuis, David Goodman, Michael Watts, Dick Walker, and Elizabeth Barham. In Italy we'd like to thank Gianluca Brunori and Claudio Cecchi. In the United Kingdom we must thank David Barling, Bill Goldsworthy, Duncan Green, Tim Lang, Bob Lee, Peter Midmore, Louis Morgan, Robin Morgan, Sue Morgan, Rory O'Sullivan, Pam Robinson, Andrew Sayer, and Neil Ward. Among our colleagues in the School of City and Regional Planning we would like to thank Janice Edwards, Andrew Flynn, Mara Miele, Selyf Morgan, Cynthia Trevett, and Diane Tustin. We also acknowledge the marvellous research assistance of our colleague Joek Roex. We owe a very special debt of gratitude to Roberta Sonnino for helping us through a very difficult period in the closing stages of the book. Her editorial contribution was second to none and it went way beyond the call of duty. We are also enormously grateful to the Economic and Social Research Council for supporting our work on local and regional food systems. Last but not least, at Oxford University Press we shall always be grateful to Anne Ashby for her patience and humanity when schedules slipped (again and again). Needless to say, none of the above is in any way responsible for the shortcomings of the book.

CONTENTS

LIST OF FIGURES

LIST OF TABLES

LIST OF ABBREVIATIONS

AFI	Alternative food initiative
AFN	Alternative food network
ANT	Actor–Network theory
AoA	Agreement on Agriculture
AOC	*Appellation d'origine contrôlée*
CAFOD	Catholic Agency for Overseas Development
CAP	Common Agricultural Policy
CCOF	California Certified Organic Farmers
COFA	California Organic Foods Act
COOL	Country of Origin Labelling
CSA	Community supported agriculture
DEFRA	Department for Environment, Food and Rural Affairs
DOC	*Denominazione d'Origine Controllata*
DOCG	Designation of Controlled and Guaranteed Origin
EPA	Environmental Protection Agency
EQIP	Environmental Quality Incentive Program
FDA	Food and Drug Administration
FSA	Food Standards Agency
FT	Fairtrade
GATT	General Agreement on Tariffs and Trade
GI	Geographical Indication
GMO	Genetically modified organism
LFA	Less Favoured Area
MAFF	Ministry of Agriculture, Fisheries and Food
MAP	Modified atmosphere packaging
NAFTA	North American Free Trade Agreement
NGO	Non-governmental organization
NIMB	Not in my body
PDO	Protected Designation of Origin
PGI	Protected Geographical Indication
PSE	Producer Support Estimate
rBGH	Recombinant Bovine Growth Hormone
RDP	Rural Development Plan
RDR	Rural Development Regulation
SDT	Special and Differential Treatment
SFSC	Short food supply chain
TSE	Total Support Estimate
USDA	United States Department of Agriculture
WTO	World Trade Organization

Introduction

When Guillermo Vargas from Costa Rica visited the British House of Commons in 2002 to publicize Fairtrade Fortnight, he delivered a stark message. 'When you buy Fairtrade', he said, 'you are supporting our democracy'. It is hard to imagine a more powerful testament to the ripple effect of our food choices. Buying food may be a private matter, but the type of food we buy, the shops or stalls from where we buy it, and the significance we attach to its provenance have enormous social consequences. Our food choice has multiple implications—for our health and well-being, for economic development at home and abroad, for the ecological integrity of the global environment, for transport systems, for the relationship between urban and rural areas and, as the Fairtrade story shows, for the very survival of democracy in poor, commodity-producing countries.

Although food consumption habits show considerable differences between countries, and between social classes within countries, a number of generic trends have emerged in recent years, some of which have been attributed to the globalization of style and taste. In the processed food cultures of the US and the UK, for example, the key trends include the increasing popularity of convenience foods, the decreasing amount of time devoted to preparing meals, the falling share of money devoted to food in the household budget, the primacy of price when buying food, and, more recently, burgeoning concerns among all classes of consumer about the quality and safety of food.

Some of these trends appear to be contradictory, particularly the emphasis on cheap food on the one hand and the growing demand for healthy food on the other. Another example might be the growing interest in local food, which is often equated with fresh and wholesome produce, and 'global sourcing', which aims to transcend the constraints of locality and seasonality. Conventional food retailers are acutely conscious of the need to accommodate these conflicting signals, as a trade body in the UK freely acknowledged when it said that 'the industry challenge is to find a balance between supporting British farmers and reducing food miles, and satisfying consumer demand for year round availability of an increased number of products, at ever lower prices' (IGD, 2002).

However, these different food trends may be less contradictory than they appear considering that, to a large extent, they reflect the food choices of different social segments of the market. These consumer patterns also correspond to very different agri-food systems. Although we try to eschew binary oppositions in this book, it is useful to draw a stylized distinction between two agri-food systems, namely: the *conventional* system, which is dominated by productivist agriculture and large companies producing, processing, and retailing food on a national and global scale, and the *alternative* system, which tends to be associated with a more ecological approach to agriculture, with smaller companies producing and retailing food for localized markets. This distinction is of course something of a caricature because, as we show later, the border between these systems is becoming more and more porous. For example, not only are conventional supermarkets increasingly interested in selling local food, but they are already the largest retailers of organic food—two categories of food that are indelibly associated with the received image of the alternative sector (Morgan and Murdoch, 2000).

From the standpoint of the conventional system, the history of agriculture is a productivist success story of the highest order. One of the proudest boasts here is that agriculture has delivered something that previous generations could only dream about, namely a ready supply of cheap food that is accessible to, and affordable by, the vast majority of people in the (western) world. Certainly on the conventional metric, that extols quantity over quality in a mass production system designed to reap economies of scale for producers and low prices for consumers and that is deeply embedded in Anglo-American corporate culture, the record looks like one success story after another, as food supply became progressively 'liberated' from nature and her seasons. Agriculture, in this conception, is just another economic sector, part of the consumer goods industry.

In recent years, however, a rival interpretation has emerged. This is based not on the productivist metric of mass production, but on the ecological metric of sustainable development, a metric that invites us to internalize the costs that are externalized in the conventional food system. The externalized costs of the conventional food system are perhaps most apparent in terms of environmental and healthcare costs. The main environmental costs are related to the global production and distribution of food. On the production side, the costs are mainly associated with the intensification of agricultural production, which has caused declining soil fertility, water pollution, animal welfare problems, and the loss of valuable habitats and landscape features (Pretty, 1998; 2002). On the distribution side, the environmental costs of food miles have been well documented. Moreover, despite the fact that aviation is the most damaging mode of transport, there is no tax on aviation fuel, a glaring anomaly from the ecological standpoint (A. Jones, 2001).

Human health is another sphere where the externalized costs of the conventional food system are becoming ever more apparent. Among nutritionists, the year 2000 was very significant because for the first time the number of overweight people in the world matched the number of undernourished people, with 1.1 billion people in each category (Nestle, 2002). The escalating financial costs of diet-related disease are placing intolerable burdens on healthcare systems, particularly in the US, where the consumption of foods high in fat, sugar, and salt is associated with high levels of obesity throughout the population. Recent scientific findings suggest that fast food creates an addictive effect not unlike that of tobacco, leading some authors to argue that obesity may be less a problem of gluttony and fecklessness, and more a problem of vulnerable human genes in a hostile food environment (E. R. Shell, 2002). Whatever the precise cause, obesity presents the conventional food system with an enormous problem—the problem of anti-obesity litigation from aggrieved consumers and cash-strapped governments. The underlying rationale of anti-obesity litigation is to make the conventional food system face up to, and pay for, the costs it has externalized on to others.

Disquiet about the health and environmental effects of the conventional food system has fuelled increasing anxiety about both food supply and the regulatory regime that is responsible for policing it. Arguably, the field of food provision has become one of the most controversial in the political arena as well as at the level of everyday life (Harvey, McMeekin, and Ward, 2004). One of the effects of the decline of public trust in the conventional food system is that certain consumers—particularly educated, middle-class consumers—are becoming ever more concerned about *where* their food comes from and *how* it is produced and distributed (Bell and Valentine, 1997). A growing sensibility to the place and provenance of food provisioning is often construed as a boon for the alternative food system, which trades on the quality attributes of authenticity and traceability. What is less often appreciated, however, is the extent to which place and provenance are insinuating themselves into the conventional food system. For example, the current political struggle in Europe and the US over food labelling policy is in part a conflict about whether consumers have the right, or even the need, to know the social and spatial history of their food. As we shall see, the corporate agribusiness sector, particularly in the US, argues that consumers have little or no interest in the place and provenance of their food, whereas consumer, health, and environmental campaigners beg to differ. Far from being an innocent technical arena, then, food labelling policy is a key site of 'the quality battleground' in the contemporary food chain (Marsden, 2004b).

The underlying themes of the book—place, power, and provenance—were chosen because they encapsulate some of the most compelling political issues in the agri-food system. *Place* has always bedevilled social and spatial theorists because of its inherent ambiguity. Like D. Harvey (1996: 208), however, we consider the 'multiple layers of meaning' an advantage, rather than a problem. These multiple meanings range from 'place' as a jurisdictional entity, such as a local authority district, to 'place' as a relational construct, where social or political relations are the determining forces, rather than formal administrative boundaries. Although the capitalist process of 'creative destruction' is ultimately what drives the making and breaking of places, this is a deeply mediated process, especially when state action is invoked to temper or resist the logic of market forces. Some of the rural places that we examine in later chapters are highly distinctive because, for much of the post-war period, they were part of a state controlled agri-food system, rather than a market regime. As this state system in Europe and the US is gradually liberalized, these rural places have to invent new vocations for themselves, for example by diversifying into quality products that play upon their association with place and provenance. Adjusting to a more spatially conscious world of production and consumption is much less of a challenge for such countries as Italy and France, where a link between places and products has been maintained, than for such countries as the US and the UK, where regionally distinctive products long ago gave way to the anonymity of manufactured products, the legacy of which is a 'placeless foodscape' (Ilbery and Kneafsey, 2000: 319).

Although it is often conflated with place, *provenance* has a much wider meaning. Its literal meaning—which is the place of origin or the earliest known history of something—is an ambiguous amalgam of the spatial and the social, of geography and history. With respect to food, we use the term in the widest sense to embrace a spatial dimension (its place of origin), a social dimension (its methods of production and distribution), and a cultural dimension (its perceived qualities and reputation). The social dimension is particularly important because it helps consumers to deal with the ethical issues in globally dispersed food supply chains, including the employment conditions of food production workers; the welfare of animals farmed as food animals, such as battery hens and veal calves for example; the integrity of some food production methods, such as adding hormones to beef for instance; the environmental effects of certain production methods, such as the use of pesticides and the destruction of flora and fauna. To the extent that a new moral economy is beginning to emerge around food issues, this question of provenance assumes a central importance in food chain regulation.

If place has multiple layers of meaning, so does *power*. Running through the manifold forms of power addressed in this book is a conception that understands power in terms of a capacity to mobilize, control, and deploy resources—be they economic, political, cultural, or indeed moral—a conception that recognizes the distinction between 'hard' and 'soft' power. The common currency of the corporate and political worlds, hard power ultimately involves the power to cajole, to compel, and to command by force if necessary. As we shall see, the exercise of hard power is a routine feature of the conventional food system, especially in retailer-led supply chains where primary producers have been so emasculated that, in some cases, the prices they receive from supermarkets can be lower than their costs of production, a manifestly unsustainable relationship. Soft power, by contrast, refers to the capacity to enlist, to inspire, and to persuade through ethical and/or intellectual argument, a form of power that is more prevalent in the alternative food system and something we explore through the prism of the new *moral economy* of food.

A moral economy perspective could significantly enrich the agri-food literature, a burgeoning but under-theorized field. While economic geography and rural sociology have both punched above their weight in the past to establish the subject we know today as 'agri-food studies', these disciplines need to be supplemented with new perspectives. As well as opening itself up to moral economy, agri-food studies could also benefit from more critical engagement with theories of multilevel governance because, far from being a local matter, food chain localization will need to draw support from every tier of the multilevel polities that govern our lives today.

The twin perspectives of moral economy and multilevel governance help to shape the analysis in the following chapters. Chapter 1 reviews some of the theoretical literature that we consider to be most relevant to the task of theorizing the 'worlds of food' that straddle the conventional and alternative food systems. In particular, we focus on the contribution of three sets of theories, namely political economy, actor–network, and conventions theory, to examine what each has to offer. Chapter 2 examines the protean regulatory world of agri-food at three different levels of governance: the global level, where we focus on WTO efforts to liberalize world agriculture; the EU and US levels, where we show that the farm support systems are being reregulated rather than deregulated; and the UK level, where we examine the advent of a dedicated Food Standards Agency to champion the neglected consumer voice in a food system hitherto dominated by producer interests. Chapter 3 extends the thematic focus by examining the changing geographies of agri-food, contrasting the deterritorializing thrust of the conventional food system with the reterritorializing logic of the alternative food system.

Following the three opening thematic chapters, we turn to consider three regional worlds of food in Tuscany, California, and Wales. We selected Tuscany because it is one of the pioneering regions in Europe for what we call 'localized quality' production, a system that aims to offer an alternative to the productivist philosophy of the conventional food system. If Tuscany is a European pioneer, then California is certainly an American pioneer, and perhaps even a global pioneer, because it is deemed by some geographers to be the world's most advanced agricultural zone (Walker, 2004). As the world's sixth largest economy, and with a state population of 34 million people, California is more akin to a European country than a European region. However, as we were less interested in the issue of comparative scale and more interested in a pioneering world of food, we decided to sacrifice the former for the sake of the latter. What is perhaps most distinctive about California from the perspective of this book is that it is playing a pioneering role in the conventional *and* the alternative food systems. After focusing on the frontier worlds of Tuscany and California, we turn to consider the peripheral world of basic commodity production in Wales. Paradoxically, the situation in which Wales finds itself is probably much closer to the majority of regions around the world than is the situation in Tuscany and California. To this extent, the Welsh Agri-food Strategy, designed to help the country escape the 'commodity world' and break into the 'quality world', may be far more instructive to other regions that are engaged in making this transition. Finally, Chapter 7 examines how the three themes of place, power, and provenance play out in the conventional and ecological food systems, blurring the boundaries between them and creating increasingly complex worlds of food.

1

Networks, Conventions, and Regions: Theorizing 'Worlds of Food'

Introduction

Food is a long-standing productive activity which carries a number of different production and consumption attributes. However, much of the recent literature focuses on a limited number of such attributes—namely, the transformation of the food chain and, more in general, of production sites. In particular, much attention has been paid to globalization, the growing power of transnational corporations and their relentless exploitation of nature.

In this chapter we argue that this kind of focus is not alone sufficient to account for the growing complexity of contemporary agri-food geography. Growing concerns about food safety and nutrition are leading many consumers in advanced capitalist countries to demand quality products that are embedded in regional ecologies and cultures. This is creating an alternative geography of food, based on ecological food chains and on a new attention to places and natures, that, as we will see in Ch. 3, reveals a very different mosaic of productivity—one that contrasts in important respects with the dominant distribution of productive activities so apparent in the global food sector (Gilg and Battershill, 1998; Ilbery and Kneafsey, 1998).

Our aim is to develop an analytical approach that can aid our understanding of this new agri-food geography and can introduce a greater appreciation of the complexity of the contemporary food sector. To this end, we begin by considering work on the globalization of the food sector and by showing that recent analyses have usefully uncovered some of the key motive forces driving this process—most notably the desire by industrial capitals both to 'outflank' the biological systems and to disembed food from a traditional regional cultural context of production and consumption. After considering the recent assertion of regionalized quality (which can be seen as a response to the outflanking manœuvres inherent in industrialization), we examine approaches such as political economy, actor–network theory, and conventions theory that have made significant in-roads into agri-food studies and have revealed differing aspects of the modern food system. In doing so, we highlight what we consider the main limitation of these approaches: i.e. their tendency to conceptualize the contemporary agri-food geography in terms of binary oppositions—such as, for example, conventional v. alternative, and global v. local.

In order to begin to overcome such binary thinking, in the last part of the chapter we analyse and expand Storper's theory of productive worlds. We feel this theory helps to engage with the varied outcomes that now exist in the contemporary food sector and can therefore highlight the implications of different productive systems on differing spaces and places. However, we also suggest that Storper's theory needs some modification if it is to be made applicable to the analysis of the contemporary food sector. In particular, we highlight two aspects that require further work: one, the key role that nature plays in the production and consumption of food; two, the activities of political institutions situated at differing levels of the polity—including regions, nation-states, and international organizations. We attempt to integrate these two features into Storper's general approach in order to conjure up differing worlds of food. The notion of worlds of food that emerges from this analysis will guide the discussion in later chapters.

A Bifurcated Food Sector?

For some time now it has been widely believed that the agri-food system is *globalized*. As a consequence, much recent research (see e.g. Goodman, 1991; Goodman and Redclift, 1991; Goodman and Watts, 1994; 1997; Goodman, Sorj, and Wilkinson, 1987; McMichael, 1994; Whatmore, 1994) has taken as its main focus how processes of globalization come to be driven by the reshaping of food production processes according to patterns of capital accumulation. In many respects, the globalization of the food system follows the same course as globalization in other economic sectors, that is, production chains are increasingly orchestrated across long distances by a few large-scale economic actors, usually transnational corporations (Dicken, 1998). In other important respects, however, the development of the food system follows its own course due to some specific characteristics of food production, notably its close association with a natural resource base and cultural variation in consumption practices (Goodman and Watts, 1994). In our view, the globalization of the food sector is uniquely constrained by nature and culture: food production requires the transformation of natural entities into edible form, while the act of eating itself is a profoundly cultural exercise, with diets and eating habits varying in line with broader cultural formations. These two key aspects necessarily tie food chains to given spatial formations. In other words, food chains never fully escape ecology and culture. Thus, in order to understand the development of the agri-food sector it is necessary to consider how forces promoting globalization interact with natures and cultures that are spatially 'fixed' in some way. In the following pages we consider the 'fixity' of nature and culture and show how the interaction between mobile and fixed resources underpins the new geography of food.

Nature

Food is necessarily a mix of the organic and the inorganic (Fine, 1994; Fine and Leopold, 1993) or the natural and the social (FitzSimmons and Goodman, 1998; Goodman, 1999; Murdoch, 1994). Thus, biology plays a crucial role in mediating social processes of industrialization and places constraints upon the extraction of profit or value from the food sector (Goodman and Redclift, 1991). In short, nature acts to localize or regionalize food production processes. Of course, to maximize productivity gains continued efforts are made by producers and manufacturers to reduce the importance of nature. We can cite just one example here: seasonality. The Italian food historian Montanari (1996: 161) emphasizes just how much producers and consumers have traditionally seen seasonality as an affliction. He says: 'symbiosis with nature and dependence upon her rhythms was once practically complete, but this is not to say that such a state of affairs was desirable; indeed, at times it was identified as a form of slavery'. This was especially true for the poorer sections of society, where consumption of foods such as grains and legumes was the norm precisely because these foods could be easily conserved. Access to fresh and perishable foods—such as vegetables, meat and fish—was the luxury of an elite few. Thus, 'the desire to overcome the seasonality of products and the dependence on nature and region was acute, though the methods for doing so were expensive (and prestigious); they required wealth and power' (p. 162). Montanari therefore concludes that it is 'doubtful whether we can attribute either a happy symbiosis with nature or an enthusiastic love for the seasonality of food to "traditional" food culture' (p. 163).

As we now know, food production processes have moved a long way from any such symbiotic state of affairs. Since the mid-nineteenth century, food has been subject to what the French food historian Flandrin (1999: 435) calls a 'never-ending Industrial Revolution'. One main purpose of this 'revolution' has been to undercut nature's restrictive powers, notably attachments to seasonality. Food preservation techniques were refined from the mid-nineteenth century onwards while new technologies such as refrigeration were introduced in the early years of the twentieth century. As the American food writer Levenstein (1999) puts it, 'producers and processors developed a host of new methods for growing, raising, preserving, precooking, and packaging foods'. The refinement of such methods intensified during the post-war period, with over four hundred new additives and preservatives developed during the 1950s alone. As a consequence, the molecular structure of food was transformed, opening the way to yet further modification in later decades (Capatti, 1999).

At the same time, the struggle against seasonality meant that food was increasingly transported over greater and greater distances in order to ensure year-round availability. The decreasing cost of transportation (the cost of sea and air transport progressively fell throughout the twentieth century—see Millstone and Lang, 2003) meant that retailers could put in place a 'permanent dietary summer' in which seasonality was forever banished. As Montanari (1996: 163) observes, the result has been that in fortunate parts of the world such as western Europe and the US 'the dream has been realised... finally we can live for the moment (just like Adam and Eve before the fall) without worrying about conserving or stockpiling. Fresh seasonal food is a luxury that only now... can be served at the tables of many'.

The example of seasonality shows that as nature is squeezed out of the production process, so global linkages are increasingly consolidated, making the food system an intrinsic part of globalized commodity production. A great deal of work in agri-food studies therefore concerns itself with how multinational companies, research and development agencies, and state actors combine to push the globalization process in the food sector in ways that ease any natural restrictions (Bonanno et al., 1994; Friedland et al., 1991; Heffernan and Constance, 1994; Lowe, Marsden, and Whatmore, 1994; Raynolds et al., 1993). Yet, while recent work on the globalization of food has concerned itself with a restructuring of the food sector in line with the demands of internationalized agri-food industries, it is also recognized that production processes are still mediated and sometimes refracted by regional and local specificities (Arce and Marsden, 1993; Goodman and Watts, 1997; Marsden and Arce, 1995; Marsden et al., 1996; Page, 1996; Ward and Almås, 1997). This local refraction of global processes seems to be intrinsic to the industrialization of the food sector, in part because the various mixtures between the organic and inorganic are hard to detach from space and place. Referring to agriculture, Page (1996: 382) says that industrialization continues to be 'conditioned by the natural basis of production, as well as by the social relations that often follow closely in the wake of natural difference, resulting in distinctive processes of economic and spatial growth'. He then argues that these peculiar features mean 'patterns of uneven development in agriculture are not solely the outcome of industrial dynamics, but are produced through the complex articulation of these processes with diverse sets of places' (p. 389). Moreover, 'embedded local conditions have important effects upon agriculture, often serving as powerful barriers to industrial transformation'.

In short, contemporary food chains are not as 'disembedded' from local natures as a superficial reading of the globalization literature might indicate. However, the role of nature should not simply be confined to that of a 'residue', one that is likely to be gradually displaced by the development of new technologies (such as genetic modification). Rather, nature displays what Beck (1992) has termed 'boomerang' qualities, that is, it has a habit of bouncing back in the wake of human modification. The most notable example of nature's boomerang quality in the food sector is Bovine Spongiform Encephalopathy (BSE), where a seeming domestication of various natural entities suddenly gave rise to a terrifying new actor (a prion protein) that causes irreversible destruction to the human brain and, ultimately, human life. As this case importantly illustrates, the food sector can attempt to 'outflank' nature but these outflanking manœuvres can bring problems in their wake (Goodman, 1999).

The health scares associated with BSE—and other illnesses such as salmonella, and *E. coli* poisoning—have resulted in an enhanced consumer sensitivity to the ways and means of food production and processing (Griffiths and Wallace, 1998). In turn, this sensitivity has put pressure on producers and processors to ensure that their foods are safe and nutritious. Again, this has tended to highlight the status of nature in food (Murdoch and Miele, 1999). Perhaps even more significantly (at least from the standpoint of the analysis being elaborated here), such pressures have promoted a re-embedding of food production processes in local contexts, in part because locally sourced food is often assumed to be of a higher quality (i.e. 'safer') than industrial (placeless) food (Nygard and Storstad, 1998). As a consequence, a sizeable and growing minority of consumers are currently turning to local and regional food products in the hope that these will offer protection against industrialization's excesses. Fernandez-Armesto (2001: 250) summarizes this trend at the end of his history of food when he says: 'an artisanal reaction is already underway. Local revulsion from pressure to accept the products of standardised taste has stimulated revivals of traditional cuisines.... In prosperous markets the emphasis is shifting from cheapness to quality, rarity and esteem for artisanal methods.... The future will be much more like the past than the pundits of futurology have foretold.' This localization of food is of course taking place in the context of globalization. Thus, we can discern a complex interaction between spatial scales as differing productive activities and products become set in varied spatial contexts. Some foods are 'global' (Mars Bars, Coca-Cola, McDonald's burgers and so on), other are 'local' (*lardo di Colonnata*, saltmarsh lamb), while yet others combine both the local and the global (Parmigiano Reggiano, Parma ham, Aberdeen Angus beef). The result is an increasingly fragmented and differentiated food market.

Culture

For some time it seemed as though the forces of standardization and industrialization would succeed in engineering a homogeneous food culture in which spatial variation became of decreasing significance. In line with this view, one commentator has recently claimed that 'the "variety" you can see on entering a supermarket is only *apparent*, since the basic components are often the same. The only difference is in packaging and in the addition of flavouring and colouring. Fresh fruit and vegetables are of standard size and colour, and the varieties on sale are very limited in number' (Boge 2001: 15; emphasis added). Yet, as we argued above, it seems that modern consumers can no longer rely so readily upon these industrialized food goods. As Beck (2001: 269) puts it, 'many things that were once considered universally certain and safe and vouched for by every conceivable authority turn...out to be deadly'. Beck suggests that, in this uncertain consumption context, many consumers become more 'reflexive' in their relationships with food and other commodities. One consequence of this more reflexive attitude is a concern for provenance, that is, the place of production. In part, as we suggested above, this is due to the fact that the ecological conditions implicated in production processes can be more easily discerned if provenance is known. Yet, there is also a cultural dimension to this; local food is likely to be produced in line with long-standing traditions, that is, by artisanal rather than industrial processes. Moreover, such foods will probably be embedded in long-standing cultures of consumption in which the qualities of the product accord with local notions of taste.

In the wake of contemporary food scares, these local cultures of consumption have been revalued. In part, the enhanced value of such cultures derives from their precarious status: they seem to be continually threatened by the diffusion of standardized and globalized food products. Moreover, these 'alternative' food cultures offer means of resisting the further standardization of food. As the Italian Slow Food organisation (*www.slowfood.com*, accessed 16 May 2005) puts it, industrialization and standardization in the food chain can best be challenged by a rediscovery of 'the richness and aromas of local cuisines.... Let us rediscover the flavours and savours of regional cooking and banish the degrading effects of Fast Food.... That is what real culture is all about: developing taste rather than demeaning it.' In other words, it is not only nature that plays a key role in safeguarding the health and nutrition of the food we eat; long-standing food cultures also play this role.

Groups such as Slow Food, which are committed to combating the standardizing impulses of globalized food chains, emphasize the need to rediscover and protect geographical diversity as a good in itself. Slow Food undertakes a whole range of activities that are aimed at strengthening markets for local and regional food products (i.e. products that have a clear connection to local systems of production and consumption—what the French call *terroir*). In this regard, Slow Food effectively voices implicit and explicit criticisms of the 'massification of taste'. These criticisms are mainly articulated *culturally*. Slow Food sees food as an important feature of the quality of life. As the Slow Food Manifesto puts it, its aim is to promulgate a new 'philosophy of taste' and its guiding principle is 'conviviality and the right to taste and pleasure'. The pleasurability of food is derived from the aesthetic and cultural aspects of production, processing, and consumption. All these activities are considered 'artful'; they require skill and care and evolve by building on the knowledges of the past to meet the new social needs of contemporary consumers.

In sum, Slow Food and other proponents of local and regional foods aim to challenge the diffusion of a fast food culture by asserting alternative cultures of food. Starting from the acknowledgement that food is imbued with symbolic meanings, and that patterns of food consumption have evolved over time according to the gradual evolution of tastes, these groups promote the values of typical products and regional cuisines because they are thought to reflect cultural 'arts of living'. As a leading Slow Food activist (Capatti, 1999: 4) puts it, 'food is a cultural heritage and should be consumed as such'. Thus, for Slow Food a cultural appreciation of food requires an appreciation of the temporal flow of food from the past into the present into the future. 'Slow food', in Capatti's view (p. 5), 'is profoundly linked to the values of the past. The preservation of typical products, the protection of species from genetic manipulation, the cultivation of memory and taste education—these are all aspects of this passion of ours for time.'

The increasing cultural value attached to local and regional foods can also be discerned in the number of new brand names or trademarks that are now appearing. In France, for instance, large numbers of agricultural products (including cheese, wine, olive oil, haricot beans, and potatoes—see Barjolle and Sylvander, 1999) are receiving the country's prestigious *appellation d'origine contrôlée* (AOC) classification a mark that reflects the local provenance and quality of the product. Following the success of the French scheme, a similar approach was adopted at the European level. In 1993 the European Community put in place legislation to protect regional and traditional foodstuffs (Council Regulation (EEC) No. 2081/92, *Official Journal* L208, 24 July 1992, p. 128). This legislation codifies definitions for products with a Protected Designation of Origin (PDO) and a Protected Geographical Indication (PGI). A PDO or PGI is defined as the name of a region or a specific place followed by the letters 'PDO/PGI', which refer to an agricultural product or foodstuff originating in that region or place. For a PDO the 'quality or other characteristics of the product are essentially or exclusively due to a particular geographical environment with its inherent natural and human components and whose production, processing, and preparation take place in the geographical area'. A PGI possesses 'a quality or reputation which may be attributed to the geographical environment with its inherent natural and/or human components'. In other words, it is the intimate intermingling of localized natures and cultures that gives PDO and PGI products their distinctive character.

In many respects, the emergence of these quality 'marks' can be seen as an attempt to tie particular qualities inherent in the *product* to particular qualities inherent in the spatial context of production (organizational, cultural, and ecological qualities). We should note, however, that the development of

AOCs, PDOs, and PGIs is highly uneven across space; while these have long existed in certain countries—France and Italy, for example—they are almost completely absent in others (by 2001 there were well over 500 products registered as PDOs and PGIs but most of these were to be found in southern Europe—between them, France, Italy, Portugal, Greece, and Spain accounted for more than 75 per cent of the total). The uneven distribution of quality certification schemes reflects the uneven distribution of surviving quality production schemes.

In the European context we then witness a significant cultural difference between the south and the north (Parrott et al., 2002). In much of southern Europe the association among *terroir*, tradition, and quality is taken as self-evident. In northern Europe, however, such associations are much weaker. For example, in the UK, with the exception of a few regional dishes (such as Yorkshire pudding, Lancashire hotpot, and Cornish pasties), there is no widespread tradition of associating foods with region of origin (Mason and Brown, 1999). British cheeses may bear place names (Cheshire, Caerphilly, etc.) but, almost without exception, these are used to describe a type of cheese, rather than its place of origin or a culture of production. Vestiges of geographical association remain for only a few products—Scottish beef and Welsh lamb, for example, have both maintained their traditional reputations as superior products. With these (and a few similar) exceptions, the UK has become a 'placeless foodscape' (Ilbery and Kneafsey, 2000: 319), dominated by nationally recognizable and homogenous brand names.

Theorizing Worlds of Food

The preceding section highlights the need to attend to the regional variations found within agri-food geography. If we concentrate our attention solely upon globalizing tendencies, we will see merely those regions that are 'hot-spots' in the globalized food production system (e.g. North Carolina's hog industry or East Anglia's grain production), places where industrialized production has become concentrated into larger and larger units (Whatmore, 1994) and where local ecologies and consumption cultures tend to reflect the standardized nature of industrial food production. The emerging concern with the 'embeddedness' of food production and consumption in regionalized nature-cultures should force us to draw another map, one which highlights those areas that have not been fully incorporated into the industrial model of production and that have retained the ecological and cultural conditions necessary for 'quality' production. However, different theories have provided different responses to this new complexity. Some have tended to argue that the emergence of new regional food cultures does little to inhibit globalization of food; others have taken these cultures more seriously. In what follows we assess a number of influential theoretical perspectives and their engagement with the new geography of food.

The Political Economy of Commodity Chains

As we have already noted, a great deal of attention has been paid in agri-food studies to processes of globalization in the food sector. Analysts have shown, in often wonderful detail, how linkages are established between differing parts of the food industry and how differing spatial areas are incorporated into those linkages. In general, the most effective theoretical approach in this endeavour has been political economy. While it is not easy to characterize the political economy approach—as it comes in a variety of forms and displays a variety of emphases,[1] it can be said that this tends to portray globalization as merely the latest stage in the development of the capitalist space economy. As Bonanno (1994: 253) puts it, 'capitalist development has abandoned its multinational phase to enter a transnational phase' in which 'the association of economic activities, identity and loyalty of conglomerates with a particular country are decreasingly visible'. In the process of this broad shift to transnationalization, agriculture and food production come to be integrated into a set of transectoral production processes.

Political economy has been widely employed to think through the consequences of this integration (see Bonanno et al., 1994; Fine, 1994; Friedland et al., 1991; McMichael, 1994; Marsden, 1988). One particularly influential variant has examined the construction of food commodity chains. The investigation of commodity chains or networks in the food sector has strong theoretical roots. The first examples of agri-food commodity chain analysis appeared during an early round of Marxist theorizing on the sector. The political economy of food chains identified an increasingly rapid destruction of traditional agricultural production forms (e.g. family farms) as the imposition of capitalist relations fuelled a process of industrialization (de Janvry, 1981). This process appeared to be 'disembedding' food production processes from pre-existing ('pre-industrial') economic, social, and spatial connections. For instance, work conducted in the United States by Friedland, Barton, and Thomas (1981) discerned differential rates of capitalist penetration in the agri-food sector (which varies, the authors argue, according to the commodity in question) but concluded that the process is well advanced across the food sector as a whole. Within each commodity chain, differing mixtures of technical, natural, and economic resources are integrated so that a number of distinctive industrial structures (of which agriculture is a diminishing part) are evident. The notion of 'commodity chain' is used because it shows how different commodity sectors are organized and highlights the complex sets

---

[1] For a useful overview, see Buttel, Larson, and Gillespie (1990).

of relationships that are necessarily invoked within each organizational segment.

The political economy of commodity chains was tailored to the sets of relations that are typically constructed around different agri-food commodities. Friedland (1984) summarizes the research foci of the early studies as: the labour process; grower and labour organizations; the organization and application of science and technology; and distribution and marketing. As this list indicates, the commodity system approach deals largely with the economic and social dimensions of industrialization (see Buttel, Larson, and Gillespie, 1990). Friedland (2001: 84) has recently admitted that commodity system analysts frequently take as their main concern 'agricultural mechanisation and its social consequences'. They therefore tend to focus upon industrial rationalization of the chains and the way this configures production relations at the local level.

More recently, another aspect of commodity chain activity, one that implies a renewed significance for the spatial distribution of resources, has come to the attention of food sector analysts; this is the environmental or natural components that are often so central to food chain construction processes (both in terms of production—e.g. seasonality, perishability, pollution—and consumption—e.g. quality, health, safety). In their early work, Friedland, Barton, and Thomas (1981) noted that the specific character of agri-food chains is often determined (at least to some extent) by the natural properties of the commodity itself (e.g. the perishability of lettuce and tomatoes). This insight is taken further by Goodman, Sorj, and Wilkinson (1987), who point out that the consolidation of capitalist enterprises in the food sector goes hand in hand with a need to replace and substitute natural processes as part of an effort progressively to squeeze biological constraints out of the production process (see also Goodman and Redclift, 1991). Goodman, Sorj and Wilkinson also argue that an expansion and lengthening of food networks tend to result from the progressive industrialization of food so that food products come to be transported over longer and longer distances. This lengthening of food chains increases their socio-technical complexity and leads to the emergence of global commodity chains.

It appeared from this work on commodity chains as though natural resources were of diminishing significance in the food sector—ultimately nature would be 'outflanked' by processes of appropriationism and substitutionism. Yet, Goodman (1999) has asserted that this prediction has proved only partially true: while technologies such as genetic modification do continue the outflanking procedures identified in earlier stages of food sector development, other trends give nature a new-found significance. This can be discerned, Goodman suggests, in the popularity of organic foods, which are held to retain key natural qualities, and in the consumption of typical and traditional foods, which are believed to carry cultural qualities long embedded in traditional cuisines. The increasing popularity of these food types, he argues, challenges the instrumental rationalities of the industrialized food sector and implies the need for more 'embedded' forms of production and consumption. They also challenge, he suggests, the significance of the political economy approach; in fact, while this theoretical repertoire has helped to render visible the new connections and relationships that surround and shape food commodities, it leaves little theoretical space to discern much deviation from the precepts of 'capitalist ordering' (either on the part of producers or consumers). In other words, it fails to appreciate the full significance of the new ecological conditions that Goodman believes exist in key parts of the contemporary food sector.

It is indeed true that political economists have been wary of attributing too much significance to the local production processes that give rise to niche products. There is a feeling among many such analysts (see Friedland, 1994) that the countervailing movement against globalization simply pales into insignificance in comparison with the huge global flows that now characterize the contemporary food sector. While growing numbers of consumers may be turning to 'alternative' food products, the vast majority can still be found in mass markets. Moreover, the GM juggernaut continues to roll and this holds the potential to unleash a further round of industrial development. Thus, we must balance any celebration of localized natures and cultures against a recognition that processes of industrialization and standardization continue to unfold.

Actor–Network Theory

In seeking to move beyond political economy, Goodman (1999) turns to actor–network theory (ANT)—an approach developed in the sociology of science and technology (but now more widely applied[2])—because he believes it shows more clearly how natural and social entities become entwined with one another in food networks. ANT authors (notably Callon, Latour, and Law) argue that chain or network activities can only be totally comprehended by taking into account the full range of entities (natural, social, technological, and so on) found therein. It is in this context that Callon (1991: 133) defines a network as 'a coordinated set of heterogeneous actors which interact more or less successfully to develop, produce, distribute and diffuse methods for generating goods and services'. This focus on 'hybridity' appears to accord better with Goodman's concern for the new 'ecology' of food.

ANT differs in important respects from the political economy approach. Whatmore and Thorne (1997: 250), for instance, see the political economy of globalization as involving a tendency to evoke 'images of an irresistible and unimpeded enclosure of the world by the relentless mass of the capitalist

---

[2] See e.g. Law and Hassard (1998).

machine'. ANT, however, 'problematises global reach, conceiving of it as a laboured, uncertain, and above all, contested process of "acting at a distance" '. In so doing, ANT aims to 'deconstruct' the power of the powerful by showing how they struggle to maintain the myriad relationships upon which their power is based (see Callon and Latour, 1981). ANT thus aims to avoid any reification of capitalist ordering processes. Moreover, where political economy tends to see a bifurcation of global and local processes—they are often thought to be distinct and unrelated—ANT uses the same framework of analysis for both long and short networks: that is, it focuses on the precise strategies that network builders use and on the amount of work required in holding alliances, associations, and relations together.

Whatmore and Thorne suggest that, by using ANT, multiple forms of agency can be given more consideration when describing the establishment of food commodity chains. In particular, they propose that food networks must be conceptualized as *composites* of the various actors that go into their making. In this view, networks are complex because they arise from interactions among differing entity types: the entities coalesce, exchange properties, and (if the network is successfully consolidated) stabilize their joint actions in line with overall network requirements (Latour, 1999). The emphasis on heterogeneity here means that, as Callon (1991: 139) puts it, 'impurity is the rule'. Whatmore and Thorne (1997: 291–2) embellish this point:

to be sure, people in particular guises and contexts act as important go-betweens, mobile agents weaving connections between distant points in the network.... But, insists [actor–network theory], there are a wealth of other agents, technological and 'natural', mobilised in the performance of social networks whose significance increases the longer and more intricate the network becomes... such as money, telephones, computers, or gene banks; objects which encode and stabilise particular socio-technological capacities and sustain patterns of connection that allow us to pass with continuity not only from the local to the global, but also from the human to the non-human.

In other words, networks and commodity chains inevitably mobilize a multiplicity of (social, natural, technological) actors, and the longer the networks and chains, the greater the mobilization is likely to be.

Instead of the simplified world of capitalist ordering, we here encounter complex arrangements that comprise multiple rationalities, interrelated in a variety of ways according to the nature and requirements of the entities assembled within the networks. This emphasis on the heterogeneous quality of network relationships does not necessarily imply, however, that each chain or network is unique (a uniqueness that is determined only by the combination of heterogeneous elements). Networks are rarely performed in radically new or innovative ways; rather, incremental changes lead to 'new variations' on 'old themes'. Because network 'orders' tend to reflect widely dispersed 'modes of ordering', we see patterns and regularities in network relationships.[3] Modes of ordering—which can be conceptualized as discursive frameworks holding together knowledge about past performances of network relations—are 'instantiated' and stabilized in given network arrangements. As Whatmore and Thorne (1997: 294) put it, networks perform multiple 'modes of ordering', which influence the way actors are enrolled and how they come to be linked with others.

The notion 'mode of ordering' as used in actor–network theory provides perhaps one means of establishing a connection between commodity chain and localized nature-cultures. However, actor–network theory tends to render ordering processes, and thus any connections to specific nature-cultures, in rather simplified terms. For instance, after uncovering a considerable amount of socio-natural complexity in food commodity chains, Whatmore and Thorne (1997) identify only two ordering modes in the food sector—one that arranges materials according to a rationality of 'enterprise' and another that emphasizes the spatial 'connectivity' of entities and resources. Given that food networks come in many shapes and sizes, this twofold typology seems unduly restrictive.

Conventions Theory

Closely allied to ANT is another theoretical approach that is becoming increasingly influential in agri-food studies (especially in France where it originates—see Allaire and Boyer, 1995; Boltanski and Thevenot, 1991; Eymard-Duvernay, 1989): that of conventions theory (see Wilkinson, 1997a; 1997b). Conventions theory proceeds from the assumption that any form of coordination in economic, political, and social life (such as that which exists in chains and networks) requires agreement of some kind among participants (as opposed to the simple imposition of power relations by one dominant party). Such agreement entails the building up of common perceptions of the structural context. Storper and Salais (1997: 16) describe such perceptions as

a set of points of reference which goes beyond the actors as individuals but which they nonetheless build and understand in the course of their actions. These points of reference for evaluating a situation and coordinating with other actors are essentially established by *conventions* between persons.... Conventions resemble 'hypotheses' formulated by persons with respect to the relationship between their actions and the actions of those on whom they must depend to realise a goal. When interactions are reproduced again and again in similar situations, and when particular courses of action have proved successful, they become incorporated in routines and we then tend to forget their initially hypothetical character. Conventions thus become an intimate part of the history incorporated in behaviours.

---

[3] For more detail on ordering processes see Law (1994).

Storper and Salais (p. 17) emphasize that points of reference are not imposed upon actors by some all-encompassing social order (as in political economy); rather, they emerge through the 'coordination of situations and the ongoing resolution of differences of interpretation into new or modified contexts of action'. Efforts at coordination give rise to conventions, defined as 'practices, routines, agreements, and their associated informal and institutional forms which bind acts together through mutual expectations' (Salais and Storper, 1992: 174).

Any food production activity will therefore give rise to a particular set of conventions as participants coordinate their behaviours and reach agreements on the most appropriate courses of economic action. Clearly, such agreements can cover any number of processes and eventualities. Thus, we might expect that conventions will come in many shapes and sizes. However, empirical studies by conventions theorists have tended to throw up only a limited number of convention types. For instance, Thevenot, Moody, and Lafaye (2000) identify the following as salient in providing modes of evaluation for productive and other activities: 'market performance', in which agreement is based on the economic value of goods and services in a competitive market; 'industrial efficiency', which leads to a coordination of behaviour in line with long-term planning, growth, investment, and infrastructure provision; 'civic equality', in which the collective welfare of *all* citizens is the evaluatory standard of behaviour; 'domestic worth', in which actions are justified by reference to local embeddedness and trust; 'inspiration', which refers to evaluations based on passion, emotion, or creativity; 'reknown' or 'public knowledge', which refers to recognition, opinion, and general social standing; and, lastly, 'green' or 'environmental' justifications, which consider the general good of the collective to be dependent upon the general good of the environment. Thevenot, Moody, and Lafaye (2000) argue that these convention types exist, in various combinations, in all social contexts. They therefore suggest that social scientific analysis should examine the way differing cultural formations weave together the differing combinations.

In applying this perspective to the food sector we need to consider how differing food cultures mobilize particular convention types and how these types are woven together into a coherent cultural framework. We also need to consider how consumption and production relations are aligned within such food cultures. We can then begin to take into consideration the mixtures of conventions that underpin commodity chains or networks and the relations these imply for regionalized nature-cultures. For instance, those chains or networks where modes of ordering reflect civic and domestic conventions will align a rather different set of materials and spatial connections compared with those based on industrial criteria (Lamont and Thevenot, 2000).

Although very useful to overcome the binary thinking that characterizes most literature on agri-food, conventions theory is not per se sufficient to capture the growing complexity of the contemporary food sector. In fact, it neglects what we consider two crucial areas of research: first, the role of nature in the more localized emerging agri-food ecologies and, specifically, how this relates to opportunities for, and obstacles to, the development of those ecologies; second, the need for localized agri-food ecologies to mobilize resources to draw on more than just local resources. To become and remain sustainable, such ecologies need to be endorsed by, and draw support from, a multilevel governance system that would allow speciality to defend the local globally. As we will explore in our regional case studies and in Ch. 7, by considering the interaction among economic form (network or chain), cultural context (the market demands of consumers), political/regulatory regime and the impacts upon local and regional ecologies, we can begin to see the extent to which food chains are embedded in or, alternatively, disembedded from particular places and spaces. In turn, this should allow us to examine the diverse regions that comprise the new geography of food as discrete worlds of food made up of distinct ensembles of conventions, practices, and institutions.

Conventions and Worlds of Production

The clustering of conventions, practices, and institutions in differing worlds of production is explicitly addressed in work conducted by Storper (see Storper, 1997; Storper and Salais, 1997). Storper is interested in new forms of regionalization and localization in the global economy. He argues (1997: 16) that the spatial connectedness of firms and industries can be explained not just in terms of proximity to raw materials and supplies of labour but also in terms of 'know-how', that is, 'non-codified traditions and ways of doing things [that are] essential to the job'. This know-how is enshrined in conventions, habits, routines, and other localized practices. It comprises the 'industrial atmosphere' of discrete regions and localities and gives these regions and localities comparative advantages in given industrial sectors. The uneven geography of economic activities reflects, then, a geography of knowledge, that is, the varied spatiality of codified and non-codified knowledge forms.

Storper develops his analysis by first identifying two main institutional expressions of these knowledge forms: on the one hand, there are sets of standardized, codified rules and norms that impose common conventions across a range of diverse contexts. In this institutional expression, standard procedures prescribe the way productive activities are undertaken, leaving little room for localized innovation and autonomy. This kind of codified knowledge underpins globalized economic forms. On the other hand, there are conventions that emerge from local, personalized, idiosyncratic sets of relations. Here, tacit knowledge and small-scale entrepreneurship come to the fore, although the impact of these is limited by the absence of scale economies. The difficulties involved in codifying this knowledge ensure its continued localization.

Thus two institutional expressions are demarcated: one is based on widely available knowledge and productive techniques; the other is embedded in very differentiated and distinct sets of production practices. However, Storper goes on to argue that this standardized/specialized distinction is cross-cut by the market orientation of the differing productive activities. Thus, we find, on the one hand, goods that are aimed at mass markets: these carry generic qualities and can be readily identified (through, for instance, branding strategies). On the other hand, we find goods that are produced for a dedicated market: these goods carry customized and clearly differentiated qualities that are only recognized by specialized groups of consumers.

By bringing together these two sets of distinctions, Storper (1997) identifies differing productive worlds. First, there is an Industrial World which combines standardized production processes with the dissemination of a generic product for a mass market (we might think here of well-known brands such as Coca-Cola and McDonald's). Second, there is the World of Intellectual Resources in which specialized production processes generate generic goods for the mass market (the most obvious example is the genetic modification of widely used food products such as soya). Third, there is the Market World, which brings standardized production technologies to bear in a dedicated consumer market (we might refer here to the so-called 'nichification' of food markets as products are increasingly differentiated using standardized technologies such as cook-chill). And fourth, there is the Interpersonal World of specialized production and dedicated products (clearly this refers to very localized and speciality food production and consumption practices, such as, for instance, those promoted by Slow Food).

These four worlds describe differing frameworks for economic action. In each we find particular bundles of conventions held together as standardized and specialized productive processes meet the demands of differing markets. As Salais and Storper (1992: 182) put it, 'each world must develop its own internal conventions of resource deployment, with respect to its suppliers, its factor markets, and its own internal structure'. Thus, on the one hand, in the Industrial World of standardized-generic production we expect conventions associated with commercialism, efficiency, and branding to be particularly significant. On the other hand, in the Interpersonal World of specialized-dedicated production we expect conventions associated with trust, local reknown, and spatial embeddedness to be more important.

Towards Worlds of Food

Presented in this fashion, it is clear that Storper's theory of productive worlds helps us to make sense of recent trends in the agri-food sector, where mass-market fragmentation (e.g. a growing Market World) now coexists with a resurgent specialized sector (e.g. a growing Interpersonal World). However, we wish to build on Storper's approach here by suggesting that the worlds of food that now comprise the contemporary food sector work not just according to an *economic* logic (as implied in Storper's approach) but also according to *cultural, ecological,* and *political/institutional* logics. That is, as we have emphasized above, the embedding of food in new productive worlds is taking place because of ecological problems in the Industrial World and the emergence of new cultures of consumption oriented to foods of local provenance and distinction. In short, the conventions that are assembled within new worlds of food cover economy, culture, polity, and ecology (as identified by Thevenot, Moody, and Lefaye (2000) in their various convention types).

Thus, in the Industrial World production processes and cultures of consumption are standardized (with perhaps McDonald's being the norm), while ecological factors are 'substituted' and 'appropriated'. In the World of Intellectual Resources many trajectories of development are possible but currently the dominant approach seems be a striving for an intensification of Industrial World trends—e.g. another round of appropriationism and substitutionism in the form of genetic modification and biotechnology. In the Market World production processes remain standardized but cultures of consumption are fragmenting and becoming increasingly differentiated so that many differing market niches now exist. Finally, in the Interpersonal World we find that production process, consumption culture, and regional ecology are closely bound together; they comprise a mosaic of sharply distinct 'mini-worlds' in which food consumption practices are sensitive to the ecologies of production—whether this be in the form of typical foods or organic foods.

So, we might ask, how do Storper's productive worlds map onto the new geography of food? First, it is easy to identify the spaces of the Industrial World. These are the intensive and productivist agricultural regions that are closely tied into the global economy (e.g. the Midwest of the United States, East Anglia in the UK, the Paris Basin in France). They also comprise the areas of standardized food manufacture and industrial food processes (as is the case, for instance, in Denmark). Second, the spaces that make up the World of Intellectual Resources can be seen in those laboratories and science parks that are taking forward the GM revolution (Monsanto's labs in Missouri, for instance). The fruits of such scientific know-how are applied ever more extensively in the form of GM agriculture (again, the American Midwest is the prime exemplar). Third, the Market World can be discerned in the use of standardized products for diversified market niches. The productivist agricultural regions are included here, along with new industrial spaces specializing in new food technologies such as cook-chill. Fourth, we come to the Interpersonal spaces of the local, regional, typical, and organic foods that provide the so-called 'alternative' sector. These alternative spaces, which, as we emphasized above, can be mainly found in those locations that have escaped the full rigours of industrialization, increasingly serve to demarcate food regions more clearly one from another. Here, diverse production and consumption cultures serve to strengthen those varied ecological conditions that give rise to organic and typical foods.

The four productive worlds do, then, help to make some sense of the new geography of food. However, we should not expect to be able to map these worlds easily onto discrete spatial areas; in fact, it is likely that differing nations, regions and localities will combine differing aspects of these worlds. So while some areas might be clearly dominated by the Industrial World of production, and others will be clearly dominated by the Interpersonal World, most will combine features of the differing worlds. Thus, in some places we may find industrialization sitting side by side with localization; in yet other spaces we may find more diverse market firms sitting side by side with high-tech intellectual firms. Spatially discrete worlds of food may then be more complex than Storper's theory suggests.

Conclusions

If we attempt to synthesize the various insights derived from all the theories we have analysed we find that: (1) we need to consider the 'will to power' operating in heavily industrialized food chains that work constantly to expand their reach and to override local ecological and cultural conditions; (2) even these industrial chains are to some extent based on biological processes, thus ensuring their 'hybrid' nature; (3) the complex and hybrid nature of food chains ensures that they work according to a number of differing logics, some of which emphasize efficiency or cost at the expense of nature and culture, while others work according to criteria that emphasize local connectedness, trust, artisanal knowledges and ecological diversity; (4) these differing 'logics' give rise to differing worlds of food in which conventions, practices, and institutions act concertedly to uphold particular trajectories of development; (5) these worlds of food gain spatial expression so that certain areas find themselves subject to a logic of industrialization, standardization, and efficiency while others find themselves subject to a logic of local belonging, cultural distinctiveness, and ecological diversity.

Taking all these points into account, we can conclude that the alternative geography of food is underpinned by conventions, practices, and institutions that vary in line with the differing logics of production and consumption outlined above. In short, we expect differing worlds of food to comprise differing mixtures of conventions and differing organizational forms (in terms of productive activity and market). Yet, we have chosen to focus here on specific worlds within the food sector associated with our case study regions. Thus, in Tuscany we expect to find an Interpersonal World in which many diverse, locally embedded products are produced for relatively customized markets. We expect this world to be bolstered by robust and supportive regional institutions. In Wales we expect to find an Industrial World that is busily trying to reinvent itself as a more diverse, ecologically and culturally distinctive, Interpersonal World. In California we expect to find an Industrial World that is moving inexorably into two other worlds at the same time: the Market World of diversified industrial products and the Interpersonal World of locally specific and ecologically embedded products. Yet, these starting assumptions may be flawed: we may find a much more mixed-up and complex reality in our case study regions than our initial thinking suggests. Be that as it may: we have merely stated here our working assumptions and explained how these derive from current social-scientific perspectives on the food sector.

2

The Regulatory World of Agri-food: Politics, Power, and Conventions

The history of agriculture in developed countries over the past seventy years is first and foremost a political history because of the intense interplay between farming and the state. Indeed, it is difficult to think of any other 'industry' which has been so comprehensively regulated by the state, over such a long period of time, as agriculture. Even neo-liberal governments in OECD countries have accepted the political compact between farming and the state on account of the 'exceptionalism' of agriculture. The rationale for its exceptional status might vary from country to country, but it invariably has something to do with one major aspect that distinguishes agriculture from all other industries: the fact that we ingest its products. In other words, the centrality of agriculture to human health is far and away the most important reason why many countries have sought to ensure a measure of food security by protecting their national farm sectors through permutations of production subsidies, price supports, and import controls—the origins of which stretch back to the 1930s in the case of the US and as far back as the nineteenth-century Corn Laws in the case, for example, of the UK.

Agricultural history can be read in a number of different ways. The most polarized readings are the productivist and the ecological interpretations. The productivist discourse, which emphasizes the phenomenal productivity gains that have been achieved since the Second World War, is essentially a story of unalloyed economic success due to a tripartite alliance of state, science, and farmers. The ecological discourse, by contrast, points not to the economic benefits of the post-war productivity miracle, but to the social and environmental costs of agricultural intensification. In the US, where intensive farming practices are most advanced, such problems as soil erosion and animal welfare were attributed to the regulatory regime operated by the United States Department of Agriculture (USDA), which actively encouraged unsustainable farming practices. Similar connections have been made in Europe, where the Common Agricultural Policy (CAP) was deemed to be the main culprit. When it began to emerge in the 1980s, the ecological critique of farm support policy was nothing less than a damning indictment of regulatory policy on both continents, calling into question the key political assumption that lay at the heart of this policy—that 'farmers had a special role as private producers of public goods, government support being the just reward for secure supplies of food, management of the countryside and economically viable rural communities' (Potter, 1998).

While farm support policies were assailed by ecological critics from within, an even more powerful challenge came from the wider international trade arena. Long excluded from trade liberalization agreements, agriculture was first incorporated into the GATT system in 1993, with the conclusion of the Uruguay Round of international trade talks. Liberalizing world agriculture, and the terms under which it occurs, is already a matter of 'high politics', and therefore a highly contested process, for developed and developing countries alike. In short, neo-liberal advocates of trade liberalization believe that it is the surest way of 'ridding agriculture of politics and subsidies', while critics contend that 'the GATT proposals would institutionalise a new regulatory system in the world economy privileging transnational firms' (McMichael, 1993).

However, if the European Union has its way, trade liberalization will be prosecuted as part of a process of reregulation, rather than plain deregulation, because the EU wants to continue to protect its farmers—not as food producers, but in their capacity as environmental stewards. At the heart of this argument is the concept of *multifunctionality*, which refers to the multiple roles of farming in the context of the European model of sustainable agriculture.

Reregulating agriculture through a combination of green subsidies and robust food standards may reflect the cultural conventions of producers, consumers, and citizens in the European Union, but it looks suspiciously like protectionism to struggling commodity producers in developing countries. Denied access to rich country markets in the West, and displaced from their domestic markets by the dumping of subsidized western foodstuffs, producers from developing countries are rightly cynical about the regulatory reforms of the developed countries. As we shall see, many developing countries want to see a *reregulation*, rather than a blanket deregulation, of agriculture. Like the EU, though for very different reasons, various developing country blocs are as interested in the 'non-trade' dimensions of liberalization as they are in the trade aspects themselves.

While the agricultural production end of the agri-food chain is being liberalized, the food consumption end is being subjected to ever more rigorous regulations to assure consumers that food is safe, healthy, traceable, and properly labelled. The most dramatic changes along these lines have occurred in the UK, largely because of a unique combination of food scares such as BSE and foot-and-mouth disease (FMD). These issues will be addressed in more detail in the following sections. First, we examine the politics of trade liberalization in agriculture by focusing on the current debates in the so-called Doha Development Round, where we inspect the chief contestants and their positions, highlighting the 'non-trade' aspects of the negotiations.

Second, we analyse the plans to reform the farm support systems in the two most important players in world agriculture, namely the EU and US. Specifically, we focus on the rival visions of agriculture and the conflicts that these have produced, such as those over hormone-treated beef and GMOs. Finally, to complement the focus of the first two sections, where we examine the regulatory context of trade and production, we turn to consider the new regulatory politics of *consumption* by discussing the trend towards a more consumer-conscious food system in the UK, where consumers and citizens are demanding more information about their food, in particular about its content, its origins, and the conditions under which it is produced. We use the UK to illustrate a new and more general trend in agri-food politics, namely the advent of consumer-oriented food standards agencies which are institutionally separate from—and unaccountable to—the production-oriented government departments of agriculture.

Liberalizing Agriculture: Free Trade versus Fair Trade in the Doha Round

The multilateral trade system that we know today was founded in 1948, when a General Agreement on Tariffs and Trade (GATT) was drawn up by a group of nine countries under the hegemonic influence of the US. This laid the basis for a series of trade negotiations, called 'trade rounds', which aimed at liberalizing world trade, beginning with tariffs on manufactured goods. Significantly, during the first seven trade rounds agriculture was effectively deemed to be a non-negotiable issue on account of its political sensitivity. This changed dramatically with the Uruguay Round, which began in 1986 and concluded in 1994 with two major innovations: the first Agreement on Agriculture (AoA), which established a process of trade liberalization, and the creation of the World Trade Organization (WTO), a permanent global organization, located in Geneva, which aimed at policing and promoting world trade among its member states. Another notable feature of the Uruguay Round was that its membership began to change in significant ways, not least because developing countries started to join the GATT in ever growing numbers, raising new concerns about trade, development, and poverty alleviation through agricultural reform.

The most important point to be made about the Uruguay Round, which helps to explain the critical reaction to the WTO around the world, is that it expanded the 'reach' of trade policy from arcane tariffs and such to a wide array of issues that used to be thought of as the proper domain of national governments. Across a wide range of highly sensitive fields—such as food safety and standardization, intellectual property rights, services, and the environment—the Uruguay Round gave the WTO authority in areas that were hitherto considered to be domestic issues. Another combustible element was the WTO's dispute settlement system, with its power to impose sanctions on sovereign governments and ensure that trade policy took precedence over other policies. The unprecedented 'reach' of the post-Uruguay system, embodied in the WTO, meant that this became a recipe for conflict as it 'brought trade policy into the living room' (Borregaard and Halle, 2001).

The conflicts that characterize WTO ministerial meetings, especially the 'Battle of Seattle' in 1999, can be understood in the context of this new trade policy. However, the political context in which the Doha Round was launched was also conditioned by two other factors. First, there was a widely held belief among developing countries that developed countries had not delivered on their Uruguay Round undertakings, particularly with respect to the removal of trade barriers in agriculture and textiles. Second, under the auspices of the UN, the developed countries had collectively committed themselves to a series of Millennium Development Goals to alleviate poverty and promote development, and the role of free (and fair) trade was deemed to be even more important than aid in meeting these goals.

Against this background, the WTO launched a new round of multilateral trade negotiations at Doha, Qatar, in November 2001, where members claimed they would 'seek' to place the needs and interests of developing countries at the heart of the agenda. The WTO claimed that the grandly dubbed 'Doha Development Agenda' would provide developing countries with an opportunity to trade their way out of poverty, a challenging task at the best of times but quite impossible without radical reform of the regulated farm support systems in the developed countries. The Doha Ministerial Declaration was very ambitious, especially with respect to agriculture, as the following words demonstrate (WTO, 2001):

Building on the work carried out to date and without prejudging the outcome of the negotiations we commit ourselves to comprehensive negotiations aimed at: substantial improvements in market access; reductions of, with a view to phasing out, all forms of export subsidies; and substantial reductions in trade-distorting domestic support. We agree that special and differential treatment for developing countries shall be an integral part of all elements of the negotiations and shall be embodied in the schedules of concessions and commitments and as appropriate in the rules and disciplines to be negotiated, so as to be operationally effective and to enable developing countries to effectively take account of their development needs, including food security and rural development. We take note of the non-trade concerns reflected in the negotiating proposals submitted by Members and confirm that non-trade concerns will be taken into account in the negotiations as provided for in the Agreement on Agriculture.

Dense, and subject to multiple interpretations, this highly nuanced text seemed to contain enough crafted ambiguity to offer something for everyone. The whole round was scheduled to be completed by 1 January 2005, but this timetable proved to be hopelessly optimistic. Every deadline was missed, and the key Ministerial meeting, held at Cancún in 2003, collapsed in disarray because of the lack of consensus between developing and developed countries. There were many factors behind the collapse of the Cancún negotiations (such as lack of negotiating time, lack of progress on cotton and trade-related aspects of intellectual property, new geopolitical alliances, and the unprecedented unity of the G20+ group of developing countries), but the principal reason was the failure to secure agreement on agricultural reform. In retrospect, however, if Cancún is remembered for anything, it will be for the fact that it was there that 'developing countries found their voice' (House of Commons, 2003*b*).

For many developing countries—a category that includes many different perspectives and positions—the overriding issue at Cancún was the refusal of developed countries to deregulate their farm support systems. The most potent and influential critics of regulated agriculture in the northern countries were the non-governmental organizations (NGOs)—especially CAFOD, Christian Aid, and Oxfam—which campaigned on behalf of developing countries of the south. In the run-up to the Cancún meeting, the head of research at Oxfam, Kevin Watkins, predicted the failure of the negotiations unless major agricultural concessions were forthcoming from the EU and the US. As he stated (Watkins, 2003):

The problem is this: each year rich countries spend over $1 billion a day supporting their agricultural producers—six times what they give in foreign aid. The EU and the US account for almost two-thirds of total spending. The subsidy fest translates into rocketing levels of output, fewer imports and the dumping of vast surpluses on world markets. Farmers in developing countries lose on several counts. Subsidised exports from rich countries undercut them in local and global markets, while high import barriers shut them out of rich country markets. Northern governments like to lecture on the merits of open markets. But success in world agriculture depends less on comparative advantage than on comparative access to subsidies—and poor countries lose every time.

Export subsidies are explicit in the EU, though not in the US, where the same function is performed by food aid and export credits. The debilitating effects of all these northern measures became more widely appreciated during the Doha Round, largely as a result of highly effective NGO research and publicity campaigns (Green and Griffith, 2002; Oxfam, 2002; 2003). Among other things, it was made clear that these debilitating effects included the devastation caused to the Jamaican dairy sector by the dumping of skimmed milk powder by the EU; the inability of India to compete with subsidized European milk products for exports to the Middle East; the devastating effects of the EU's subsidized sugar regime on South African small farmers—many of whom were forced out of production; the deleterious impacts of heavily subsidized US cotton exports on cotton producers in Benin, Burkina Faso, Chad, and Mali; and the inability of Ethiopian producers to compete with subsidized US corn exports to Yemen (House of Commons, 2003*a*).

Export subsidies may be the most egregious example of a trade-distorting subsidy, but they are only one of three forms of farm support identified in the AoA; the other two are domestic support and market access. Since the EU has made the most extensive use of export subsidies, it has attracted the most criticism, especially for the external effects of the CAP, which has become a potent symbol of unfair trade in the eyes of developing countries and their NGO champions. Nothing symbolized the polarized attitudes to agriculture in the run-up to Cancún better than the highly divergent positions on the CAP. While CAFOD took the view that the CAP was nothing less than 'a crime against humanity' (CAFOD, 2003), a group of seven protectionist agriculture ministers in the EU maintained that the 'CAP is something we can be proud of' (Boden et al., 2002).

The biggest problem of all in the Cancún agricultural negotiations was clearly the enormous gulf in perceptions regarding the extent to which the CAP has actually been reformed. While the European Commission genuinely believed that the CAP reform package of 2003 constituted a radical reform of the subsidy system, not least because it 'decoupled' subsidies from production and rendered them less trade-distorting, the developing countries begged to differ. The reason why these different perceptions could coexist is that the AoA provides some flexibility in the allocation of subsidies to different 'boxes'. In the arcane terminology of the WTO, domestic support falls into three categories—corresponding to amber, blue, and green boxes—according to its potential to distort agricultural trade:

- Amber Box measures include most domestic support measures that are considered to distort production and trade. These measures have to be reduced, with some WTO members pushing for their complete abolition.
- Blue Box measures include production subsidies and, therefore, they represent an exception to the general rule that all subsidies linked to production must be reduced or kept within defined minimal levels.
- Green Box measures include funds for research, the promotion of food security stocks, direct payments to producers that are decoupled from current prices or production levels, structural adjustment assistance, environmental programmes and regional assistance measures. Such measures should either have no trade-distorting effects in agricultural markets or, at the very worst, their effects must be minimally trade-distorting. The amount of Green Box subsidies is currently unlimited and no reduction commitments are required.   (WTO, 2002)

The core of the EU position on agricultural reform in the Doha Round rests on two key arguments: (1) that it must safeguard the 'European model of agriculture' because the latter delivers benefits over and above food production, as we will discuss later; and (2) that the CAP reform package, agreed in June 2003, rendered the European farm support system more WTO-compliant because the decoupling of subsidies from production meant a large shift of funding to the Green Box. This position was personified by Franz Fischler, who held the post of EU Agriculture Commissioner during the turbulent 1994–2004 period. Long before the Doha Round was launched, he succinctly summarized the EU philosophy by saying that 'the future of the European model of agriculture, as an economic sector and as a basis for sustainable development, is of fundamental importance because the multifunctional nature of Europe's agriculture and the part agriculture plays in the economy, the environment and landscape as well as for society' (Fischler, 1999).

Although the EU fought hard to get 'multifunctionality' inserted into WTO parlance, partly to have the concept externally validated, it had to settle instead for a more generic and less controversial reference to 'non-trade concerns', which means different things to different member states. To some WTO member states the expression refers to food security issues; to others it concerns the fate of fragile rural areas; in the EU, on the other hand, it tends to be a surrogate for the concept of 'multifunctionality', which highlights the value of conserving environmental benefits (such as biodiversity) and economically marginal farming systems (such as those based on the small family farm) which are deemed to be necessary for the continued joint production of food and environmental goods. More controversially, multifunctionality also implies a continued need for farm support policies to maintain the incomes of marginal farmers who are engaged in this joint production process (Potter and Burney, 2002).

'Non-trade concerns' also embrace the vexed question of animal welfare. Although there is nothing in the WTO rules to prevent a country from unilaterally raising its standards, this would have two negative consequences. First, the cost of the product would increase, leaving domestic producers exposed to cheaper imports from countries with less benign standards. Second, making a distinction among products on the basis of their production processes is highly constrained by WTO rules which prevent labelling on process and production methods, thereby making a free-range egg look much the same as an intensively produced egg. On the basis of current trends, the Royal Society for the Prevention of Cruelty to Animals forecasts that a dozen eggs in 2012 will cost 73 cents to produce in the EU, compared with 42 cents in the US. After standardizing for other inputs, the amount attributable to animal welfare standards was estimated to represent as much as 20 cents per dozen eggs, some two-thirds of the total difference (RSPCA, 2002).

Another important 'non-trade concern' for the EU relates to its campaign to secure global acceptance for its Geographical Indications (GI) system, the aim of which is to afford some protection to high-quality regional brands of food and drink, such as, for example, Parma ham, Burgundy wine, and

Roquefort cheese. A total of some 600 foods and 4,000 wines have secured protected status and the EU wants the WTO to accept and extend this system. In the run-up to the Cancún meeting, the European Commission (2003) said it would be

pushing hard for tougher rules to protect high quality and regional products, a position we are confident will be supported by other exporting countries with the same interests. The EU argues that consumer demand for specific products from specific regions provides sound business opportunities for producers all over the world. But to ensure that producers and consumers get a fair deal, these products will need to be protected against usurpation.... There is a major difference with trademarks or branded goods, which can be sold and delocalised. Not the geographical indication. The trademark is an exclusive individual right. The geographical indication is accessible to any producer of the locality or region concerned.

In the aftermath of Cancún, the EU suffered a serious setback on this issue when a WTO dispute panel ruled that the GI system was incompatible with trade rules because it did not allow the registration of non-European products. The dispute panel found that the EU could not prevent growers of Idaho potatoes or Florida oranges from protecting their food names in Europe simply because the US had not adopted a GI system. The ruling was a major victory for the US and Australia, which argued that many of the names the EU wants to protect are now generic and that the whole GI system is 'simply a new form of protectionism for Europe's already cosseted farmers' (Alden, Buck, and Williams, 2004). If the ruling is upheld, the whole European system will have to be opened up to non-European products.

To ensure fair trade for *European* producers, the EU has tabled a number of proposals at the WTO, including mandatory labelling, multilateral agreements, and welfare-friendly subsidies that enjoy the protection of Green Box status. Clearly the big challenge for the EU, and for the WTO system as a whole, is how to design a regulatory regime that promotes a liberalized agriculture without sacrificing high animal welfare standards or low-input farming systems such as organic production. What renders this task so difficult, however, is the lack of sympathy for the EU cause among the majority of WTO members, especially among cheap poultry exporters such as Brazil and Thailand, which waged a highly effective campaign in Seattle in 1999 based on the emotive slogan that the EU 'cares more about animal welfare than human welfare'—a charge that continues to resonate in many WTO and NGO circles. The EU's position in the Doha Round is not easy to categorize because, while it has less and less in common with diehard protectionists such as Norway, Japan, and Korea, it is even further removed from the free-traders in the Cairns group of countries. The Cairns group, in which Australia is one of the leading players, is highly cynical about the 'non-trade concerns' issue, believing it to be a new term for an old habit, namely European protectionism. Striking a middle way between these rival camps, one of Franz Fischler's trade negotiators described the EU's position in the Doha Round as being 'at the liberal end of the non-trade concerns club' (D. Roberts, 2002). The US position is also difficult to categorize because, while it pays homage to the free-trade ideology of the Cairns group, as we shall see in the next section, it also supports domestic farm policies that are completely at odds with this ideological stance.

Cancún changed the dynamics of the Doha Round because the 'Quad' bloc of countries, comprising the EU, the US, Canada, and Japan, finally lost some of its capacity to stage-manage the liberalization process in the face of new geopolitical alliances, such as the G20+ group of developing countries, which includes the likes of China, Brazil, India, and South Africa. Indeed, if Doha is to live up to its name as a 'development round' this will depend on developing countries taking the initiative and acting in concert, as they did at Cancún, rather than waiting for the developed countries to deliver on their undertakings, a strategy that left much to be desired in the aftermath of the Uruguay Round. In this new political environment, developed countries will have to concede far more than they originally intended, and in some cases these concessions will have to be made unilaterally, that is, without seeking some reciprocal concession from the poorest and least developed countries, which is the traditional WTO bargaining ethos. Most members agree that there will be no Doha agreement without agreement on agriculture.

Essentially, an agreement on agriculture will have to strike a judicious balance between free trade and fair trade. This will always be a provisional exercise because the tension between the two is an intrinsic and never-ending part of the contested process of international trade. Fair trade, however it is defined, will clearly need to outlaw dumping and facilitate better access to developed country markets, though border tariffs are easier to resolve than the non-tariff barriers associated with food safety standards. The latter have evolved in the EU in response to domestic pressure for more robust food safety regulations, so while they may constitute a de facto trade barrier, they were not designed to act as such, and they cannot simply be outlawed by the WTO, considering that they reflect the 'cultural preferences' of consumers and citizens. As we will see in the next section, different cultural preferences lie at the heart of two of the biggest disputes between the EU and the US— namely, hormone-treated beef and GM foodstuffs.

With regard to the developing countries, better market access will do little or nothing of itself to help them meet the exacting technical standards associated with different 'cultural preferences'. This highlights the need to align trade policy with development policy, instead of over-selling trade liberalization as a recipe for poverty reduction. Contrary to fashionable claims about the benign effects of liberalization and globalization, a sober assessment of the evidence suggests that 'integration with the world economy is an outcome, not a prerequisite, of a successful growth strategy' (Rodrik, 2001).

There can be no agreement on agriculture unless the 'non-trade concerns' of all parties are recognized and respected. For developing countries, one of the main issues in this respect is the demand for a Development Box under the WTO provision for Special and Differential Treatment (SDT), which was designed under the GATT to acknowledge a major shortcoming of universal trade rules: the fact that they treat unequals equally. The SDT concept needs to be strengthened because it is one of the key mechanisms to balance the universality of multilateral trade rules with the asymmetry of uneven development. According to CAFOD (2003), a Development Box would signal a fundamental shift in the way trade rules are designed, in that it would place food security and development needs, particularly those of poor farmers, at the heart of the negotiating process.

The idea of a Development Box, which would help developing countries to become more self-sufficient in meeting their food needs, has been strenuously resisted by the US and other agri-exporting countries. It is not the idea of food security that the US objects to, but the notion that developing countries should seek to meet their own needs from their own resources. As the US Agriculture Secretary famously declared (Bello, 2000) at the start of the Uruguay Round in 1986: 'the idea that developing countries should feed themselves is an anachronism from a bygone era. They could better ensure their food security by relying on US agricultural products, which are available, in most cases at much lower cost.' A Development Box is clearly a challenge to the system of 'agribusiness imperialism', in which the US is seeking to become a 'breadbasket of the world' through the global reach of its agri-food multinationals (McMichael, 1998). This may help to explain why the debate about a Development Box has been channelled into a more restricted proposal for developing countries to have greater flexibility on a number of 'special products' that are crucial to food security and rural development. Yet the developed countries have sought to circumscribe even this more modest proposal, a fact which shows that the notion of a 'development round' may be more apparent than real.

Developed countries also have their 'non-trade concerns' and these are especially important to the EU; indeed, they underwrite the 'European model of agriculture' and the concept of multifunctionality that lies at its heart. By transferring subsidies from the Amber and Blue Boxes into the Green Box, the EU plans to put the CAP on a more sustainable footing, rendering it compliant with multilateral trade rules on one side and meeting the multifunctional demands of European society on the other. However, the Green Box itself is now coming under pressure, with groups as diverse as the Cairns group and the G20+ group arguing that these subsidies are trade-distorting and they should therefore be capped or reduced. Since capping the Green Box would seriously constrain the 'European model of agriculture', it seems unthinkable that the EU could ever agree to such a proposal.

In short, the WTO is essentially a bargaining system in which countries trade concessions in one domain for gains in another. In the agriculture domain, all WTO members have their sacred cows and many of these fall into the category of 'non-trade concerns'. Food security in the developing countries will clearly have to be accommodated in the final agreement, which will also have to include the Green Box measures which underpin the social and spatial structure of the 'European model'.

Rival Visions: The 'European Model' versus the 'Global Breadbasket'

The leviathan farm support systems in Europe and the US are in the throes of the most radical restructuring since they were created. As we saw in the last section, the EU and the US together account for nearly two-thirds of all farm subsidies in the OECD. While each accuses the other of being the more profligate with trade-distorting subsidies, the developing countries understandably consider the EU and the US as equally culpable in this respect, so much so that they have been likened to 'two bald men squabbling over a comb' (House of Commons, 2003a). Despite all their superficial similarities, however, there are two emerging differences between the European and American systems that are potentially quite significant:

- the EU is trying to reform its farm support system with a view to creating a 'sustainable agriculture' that is embedded in a diverse and multifunctional spatial landscape, whereas the US remains more politically committed to an 'intensive agriculture' that is less attached to place and more easily delocalized in and beyond its national borders;
- the EU does not seek to impose its system of rules and regulations on other countries, whereas the US is increasingly inclined to make the world in its own image.

These different trajectories have been shaped by a series of factors, including the political influence of the corporate agri-business lobby in the US, tempered by a stronger environmental movement in Europe, at least in the northern countries of the EU. The architects of the multifunctional European model have always been acutely conscious of the different social and cultural priorities that inform agri-food policy on the two continents. 'An agriculture on the model of the US', proclaimed one EU report (European Commission, 1987), 'with vast spaces of land and few farmers, is neither possible nor desirable in European conditions in which the basic concept remains the family farm.' While the family farm is under pressure in both systems, as reflected by growing consolidation trends, this process seems to have reached a point of no return in the US, where family farming has ceased to be an official occupation because the number employed in the sector has fallen below 1 per cent of the total population—a level the US Census Bureau considers insignificant (ART, 2003).

Although the EU and US farm sectors had a broadly comparable farm gate value at the turn of the century (respectively, $197 billion and $190 billion), there are some very significant social and spatial differences in the structure of their farm systems, as Table 2.1 shows. The key differences can be summarized as follows: the farmland of the fifteen countries of the EU totalled only one-third as much as the farmland of the US, though it was three times as productive; there were over 7 million farms in the EU, compared with just over 2 million in the US; and the average size of holding was just 18 hectares in the EU, compared with 207 hectares in the US. The Producer Support Estimate (PSE), which is the main OECD indicator to assess the level of support, reached $90 billion in the EU in 2000 compared with $49 billion in the US, giving a full-time farmer equivalent support payment of $14,000 and $20,000 respectively. A more comprehensive measure of farm support is the Total Support Estimate (TSE), which includes items such as food aid which are more extensively used in the US but are excluded from the PSE calculation. On the TSE indicator, the total level of support rises to $103.5 billion for the EU in 2000, compared with $92.3 billion in the US, which means that the cost per capita was $276 in the EU against $338 in the US.

These figures can be dissected in many different ways, and they have been used to support tendentious arguments in both the EU and the US. However, the most important point to make is that the sums involved here are truly enormous, with significant opportunity costs in each case. The scale of European and American farm support underlines the perennial complaint of the developing countries: that the global geography of agri-business is shaped less by comparative advantage in world markets and more by comparative access to state subsidies (Watkins, 2003). In both regions, the vast bulk of these subsidies have gone not to small, poor farmers but to their

Table 2.1. EU and US farm support systems in comparative perspective

	US	EU
Production (farm gate value)	$190 bn	$197 bn
Number of farms ('holdings'; 1996)	2,058,000	7,370,000
Farmland	425 m. ha	134 m. ha
Average size of holding	207 ha	18 ha
Total Support Estimate (2000)	$92.3 bn	$103.5 bn
TSE per capita	$338	$276
TSE as per cent GDP	0.92 per cent	1.32 per cent
Producer Support Estimate (2000)	$49.0 bn	$90.2 bn
PSE per full-time farmer equivalent	$20,000	$14,000

Source: Eurostat figures; OECD figures for 2000.

larger and richer counterparts. For example, a comparative study of 'who benefits' found that just 10 per cent of farms in the US received 61 per cent of all subsidy payments, an uneven pattern that was replicated, albeit in a less extreme form, in the EU, where 17 per cent of farms received 50 per cent of the total. Not surprisingly, the overall conclusion of the study was that in both cases farm support policy increased income inequalities because 'the largest transfers go to a group with higher than average incomes' (Podbury, 2000).

The longevity of the CAP is a testament to the awesome power of vested producer interests in European agriculture. When the CAP was designed, in the late 1950s, the primary objective was, as the Treaty of Rome put it, 'to increase agricultural productivity by promoting technical progress' so as to achieve food security during an era when severe food shortages were still a recent memory in Europe. Forty years later the CAP found itself in a radically different context. Abundance had replaced shortage as the problem, and the producer lobby was rapidly losing out to a diverse combination of consumer, environmental, and public health groups, each calling for a more benign form of agriculture in which the emphasis was on quality rather than quantity, sustainability rather than productivity. The European Commission was becoming increasingly aware of the defects of the CAP as a regulatory regime, in particular the way in which production-related aid encouraged the intensification of agricultural techniques; how this in turn led to overproduction and a costly build-up of stocks; and the fact that income subsidies concentrated 'the greater part of support on the largest and most intensive farmers' (European Commission, 1991).

Although the CAP had survived criticisms in the past, a new combination of internal and external pressures after 2000 made reform more likely than ever before: internally there was the looming prospect of EU enlargement, which threatened to bankrupt the system; externally, it was clear that the EU would have remained on the defensive in the Doha trade talks if the CAP remained unchanged.

Like the Whig interpretation of history, the official version of the CAP tells of a series of progressively more ambitious reforms in the 1990s, such as the MacSharry reforms of 1992, which cut support prices and replaced them with direct payments to farmers, and the Agenda 2000 reforms, which introduced the Rural Development Regulation (RDR) and allowed for funds to be moved from direct support for agriculture (Pillar One) to rural development (Pillar Two)—a process called 'modulation'. These reforms are portrayed as precursors to the great reform package of 2003, which supposedly transformed the CAP into a consumer-conscious, environment-friendly, and WTO-compliant system (European Commission, 2003).

Alongside this official interpretation stands the ecological version of events, which suggests that CAP reform has been a painfully slow and highly contested process on account of the stalling tactics of a conservative alliance of member states led by France, the biggest beneficiary of CAP subsidies.

More to the point, the ecological story also suggests that the belated introduction of agri-environmental programmes does not necessarily imply a 'greening' of the CAP because green credentials were often 'a cover for the pursuit of more traditional policy goals like the support of farmers' incomes and the control of overproduction' (Potter, 1998). Nonetheless, the ecological story acknowledges the progressive trend of the 1999 reforms, especially the innovative RDR, with its capacity to accelerate the greening of the CAP. However, on the whole the 1999 reforms were judged a missed opportunity, not least because the RDR, the so-called second pillar of the CAP, was allocated just 10 per cent of funds, with the other 90 per cent still going into the first pillar of agricultural subsidies (Lowe, Bulles, and Ward, 2002; Ward and Falconer, 1999).

In the history of the CAP there is no doubt that the reform package of 2003 was the most radical since its inception. Although the final agreement fell short of what the European Commission wanted, since the original plan had to be watered down to placate France and its allies, the Commission still managed to deliver a more significant reform package than many member states thought either possible or desirable. At the heart of the package were four major innovations:

- *Decoupling*: a Single Farm Payment (SFP) took effect from January 2005 and, with minor exceptions, it is supposed to sever the link between support and production.
- *Cross-compliance*: receipt of the SFP is conditional on meeting a number of statutory environmental, food safety, plant health, and animal welfare standards, all quality criteria that were for the first time factored into the CAP support system.
- *Rural Development Regulation*: the RDR was enhanced through the injection of new resources and the addition of new schemes to promote the environment, animal welfare, and new production standards and to introduce new incentives for farmers to raise the quality of agricultural products and provide added assurance to consumers.
- *Compulsory modulation*: hitherto practised on a voluntary basis, and largely confined to the UK, modulation now became compulsory for all member states. Beginning in 2005 at a rate of 3 per cent, it will increase to 5 per cent from 2007 onwards, when it will generate an extra (1.2 billion a year for the second pillar of the CAP (*Agra Europe*, 2003a).

Buried in the small print of the 2003 reform package was a bewildering array of options that left some CAP experts wondering whether this signalled the end of a uniform EU-wide regulatory regime as 'it is now more likely than not that within a few years, no two member states will be operating exactly the same agricultural aid payment system' (*Agra Europe*, 2003b). The fact that member states can utilize a multiplicity of options helps to resolve the mystery as to how France and the UK, the two extreme ends of the spectrum on agricultural reform, could each find something positive to say about the 2003 package, with the former claiming that the CAP had been 'preserved fundamentally intact', and the latter welcoming the 'root and branch reform' (ibid.).

The 2003 CAP reform package might have been more widely welcomed had it not been for two factors, neither of which was actually part of the package. The first was the 'shoddy deal' between France and Germany in 2002, which fixed overall CAP spending until 2013. This gave the French what they wanted on agriculture, though it was more difficult to discern what Germany, the biggest net contributor to the EU, got out of it in return (*Financial Times*, 2002). The second was the fact that the reform package did not directly address some of the most controversial aspects of the CAP system, such as the sugar regime, which imposed enormous costs on poor countries such as Ethiopia, Malawi, and Mozambique (Oxfam, 2004). In this regard, the EU argued that the 2003 reform package was a process, rather than a discrete event, and that the regulatory regime for sugar would also be reformed to render it less trade-distorting.

For all the shortcomings of the reforms since 1992, it is absolutely clear that the environmental dimension of the CAP has assumed progressively more prominence in each round and that successive reforms have tended 'to consolidate the public goods model of supporting farmers for environmentally beneficial land management practices' (Ward, 2003). The fact that these reforms were driven less by environmental motives than by prosaic budgetary and trade pressures does not devalue the fact that the CAP is now addressing the issue of quality—with respect to the environment, the consumer, and animal welfare in particular—in a serious way. Tethering itself to quality of life issues, and becoming a more benign conduit for the delivery of public goods, is the only way in which the CAP can garner enough political support to survive in the twenty-first century. To reinvent itself on a sustainable basis the CAP will have to fulfil an internal role and meet an external challenge: internally, it will have to prove that the European model of agriculture is capable of delivering a multiple dividend; externally, it will have to demonstrate to the WTO that the multifunctional European model is not another name for trade protectionism.

Despite its limits as a regulatory regime, one can say that the CAP is moving in the right direction. This is more than can be said for the US farm regime. With its origins in the Great Depression of the 1930s, the US farm support system has been regulated through a series of Farm Bills, the most recent of which was the Farm Security and Rural Investment Act of 2002, commonly known as the 2002 Farm Bill. This was one of the most controversial bills of the Bush presidency because, apart from being the most expensive ever, costing some $190 billion over a ten-year period, it signalled a major reversal of the liberal thrust of farm reforms of the previous fifteen years.

What makes the US position on Doha round difficult to categorize is the fact there is no such thing as a single stance: for example, in international forums such as the WTO, the US strikes a neo-liberal posture, calling for an end to farm subsidies and the liberalization of agricultural trade, while at home it licenses the biggest farm subsidy programme in history, using the very same trade-distorting subsidies that it decries abroad. The neo-liberal *Financial Times* was outraged by the Farm Bill's 'grotesque farm subsidies' and it accused Washington of having 'surrendered to protectionism', while the heads of the WTO, World Bank, and the IMF penned a joint protest letter asking, 'how can leaders in developing countries or in any capital argue for more open economies if leadership in this area is not forthcoming from wealthy nations?' (Sumner, 2003).

This was not the first time that the US had lurched from one end of the regulatory spectrum to the other. It was largely at the behest of the US that agriculture was excluded from the GATT remit in the 1950s, a stance designed to protect its domestic farm sector from import competition. By the 1980s, however, the US had jettisoned this protectionist stance in favour of free trade, a position it also championed in the Uruguay Round. A number of factors favoured this shift. To begin with, the corporate structure of the US farm economy had changed. As we will see in Ch. 3, the most dramatic trend in this respect was the decline of the family farm and the growing concentration of the agri-food sector, which was becoming increasingly controlled by giant agri-business companies. By 1994, for example, half of all agricultural produce in the US came from just 2 per cent of farms, and the likes of ConAgra, IBP, and Cargill bestrode the sector like colossi. Having outgrown their home base, and harbouring ever more ambitious global aspirations, these giant agri-business firms had no interest in a regulatory regime that preached protectionism at home and invited others to reciprocate abroad. To this end the US began to pursue a 'global breadbasket' strategy, through which American firms would feed the world, and it sought to inscribe this corporate vision in the liberal rules and regulations of the GATT. The fact that the original US proposal to the Uruguay Round was actually drafted by Cargill's former vice-president was perhaps the most conspicuous example of the close interplay between US state policy and the corporate agri-food lobby (McMichael, 1998).

Deregulating the farm support system became the holy grail of neo-liberal reformers in the 1980s, when domestic reform in the US was perceived to be not just a precursor to, but also a prerequisite for, global reform. Prominent neo-liberals even asked whether there was anything specifically 'American' about domestic US agricultural policy, raising the spectre that regulation was an alien and aberrant philosophy in a country ideologically disposed to free trade (Paarlberg, 1989). More to the point, they reasoned that the US could not credibly seize the offensive in the forthcoming world trade talks if it had not put its own house in order. The high point of this deregulatory approach was the Federal Agricultural Improvement and Reform (FAIR) Act of 1996, which was dubbed 'Freedom to Farm' by its proponents because it removed the pre-existing system of production controls and price supports, allowing farmers to grow whatever they liked. Designed to 'get government out of agriculture', the FAIR Act offered farmers fixed but declining payments that were to be phased out completely during the life of the Farm Bill. The idea here was that increased market returns from higher exports and higher commodity prices would render government support unnecessary. In reality, however, 'Freedom to Farm' quickly became known as 'Freedom to Fail'. In fact, following the passage of the FAIR Act, exports fell and commodity prices collapsed. As a result, Congress had to abandon its deregulatory ambitions in 1998, when it launched annual emergency aid packages, making a mockery of the neo-liberal assumptions at the heart of the FAIR experiment (Lilliston and Ritchie, 2000).

The reversal of US farm policy occurred in 1998, when ad hoc emergency aid first began to be dispensed, and not in 2002, when the new Farm Bill was passed (Sumner, 2003). Although the 2002 Farm Bill is generally thought to mark the watershed in US regulatory policy, in reality it merely formalized what had been happening for the previous four years. To this extent, the bill simply recalibrated rhetoric and reality by giving the FAIR Act an official burial.

Though very different in their underlying regulatory philosophies, according to the US family farm movement (NFFC, 2002) the 1996 and 2002 Farm Bills had one thing in common: in both cases, the primary beneficiary was corporate agri-business. 'The inequities of the 1996 Farm Bill will be compounded. Those farmers (and absentee landlords) that had been getting large payments will continue to receive them, while those left out of the system will fall further behind. Ten per cent of the nation's largest farms will receive 60 per cent of the payments while the lowest 50 per cent receive little or no payments.' Stung by the fact that some 72,000 US family farms had disappeared in the 1990s, the family farming lobby tried to get four major concerns addressed by the 2002 Farm Bill: (1) a ban on meat-packer ownership of livestock to reduce the packers' power to manipulate producers and control the supply chain; (2) genuine payment limitations on the amount of subsidies flowing to large farms; (3) a family-focused Environmental Quality Incentive Program (EQIP); and (4) fairer prices from the corporate buyers of their products, rather than subsidies from the taxpayer. None of these concerns was successfully taken up. Although family farmers welcomed some green aspects of the 2002 package—particularly the new Conservation Security Program, the first ever 'green stewardship' payment scheme, and the mandatory Country of Origin labelling scheme to enhance the traceability of fish, meats, and produce—they were critical of other conservation measures. Their main criticisms were directed at EQIP, which accounted for half of all conservation funds, because it was amended to allow an individual operator (including corporate confinement animal feeding operations) to receive the highest payments over the period (NFFC, 2002). From the standpoint of sustainable agriculture, then, the 2002 Farm Bill had a whole series of shortcomings: socially, large agri-business corporations were left with free rein to force farmers into signing onerous and inequitable production contracts; environmentally, the indiscriminate use of conservation funds under EQIP would exacerbate the 'continued destruction of the rural environment' (Ikerd, 2002). Indeed, the corruption of 'green regulations' by the corporate agri-business sector was nothing new. In 1997, under pressure from agrichemical and biotechnology firms, the USDA even sought to redefine organic food standards so that the proposed 'organic' label would have allowed the use of genetic engineering, nuclear irradiation, toxic sewage sludge, and intensive animal farming practices, such as the use of antibiotics and cruel confinement conditions. This move was eventually defeated by widespread opposition from organic consumer organizations (Lilliston and Cummins, 1998).

Even though the deregulatory thrust had failed at home, the US remained singularly committed to the cause of farm liberalization abroad, and the 2002 Farm Bill did nothing to alter its negotiating stance in the WTO. On the contrary, the US evinced a more aggressive commitment to using the WTO, particularly the dispute settlement mechanism, to secure its interests through legal means rather than negotiation. Indeed, in its first five years the mechanism was largely invoked for disputes between the US and the EU, many of which concerned rival visions of agri-food regulation, such as those that emerged during the celebrated cases of hormone-treated beef and milk and genetically modified (GM) foods. These disputes merit special attention because they are profoundly emblematic of the way in which the WTO is seeking to extend its dominion into the domestic realms of countries which have hitherto been shaped by the cultural conventions of electorates.

The origins of the hormone dispute can be traced back to 1989, when the EU imposed a ban on hormone-treated beef and milk on health and safety grounds. In response to strong lobbying by Monsanto and the US beef and milk associations, Washington sought redress through the WTO. In 1997 a WTO dispute settlement panel ruled that the ban was illegal because the EU had failed to produce sufficient scientific evidence to prove that hormone-treated beef posed a risk to consumers' health. To satisfy the WTO's demand for more evidence, an EU scientific committee on veterinary medicine examined the issue afresh and concluded that at least one of the growth hormones involved, namely 17 beta-œstradiol, 'has an inherent risk of causing cancer' (Hines, 2000). On the basis of the new scientific evidence, the EU refused to lift its ban and, as a result, it incurred punitive duties. The wider point at stake here was the different philosophies of risk management, with the EU appealing to the 'precautionary principle' and the US dismissing this as 'faulty science'. Just as important as the scientific dispute, however, is the fact that EU consumers seem to support the ban.

The hormone-treated beef dispute has been dwarfed by the burgeoning GM foods dispute, a conflict that could become the most contentious trade battle between the EU and the US. The stakes are very high here; in fact, as the president of the US Grains Council put it, the application of GM technology is the 'hottest issue in agriculture today' (Hines, 2000). The economic stakes are especially high for the US because its biotechnology companies, which have invested heavily in GM technology, feel that their investments could be seriously threatened by the moratorium imposed by the EU on GM crops in 1999. When it lodged its complaint with the WTO, however, the US government went beyond economics to issue a more general warning. In fact, it saw 'the GM moratorium not as an isolated case but as symptomatic of a growing EU tendency to use health and safety as a pretext for regulations that create trade barriers' (de Jonquieres et al., 2003).

The GM dispute seems to encapsulate everything that differentiates the EU and the US in the agri-food arena—regulatory philosophy, public trust in science, and cultural attitudes to food. One of the key principles for regulating biotechnology in the US is the principle of 'substantial equivalence', which means that GM foods are not considered inherently different from other foods. US regulators, including the USDA, the Food and Drug Administration (FDA), and the Environmental Protection Agency (EPA), are obliged to base their decisions on scientific evidence that can withstand legal challenge, an approach known as 'sound science' by its proponents. The regulatory approach to biotechnology is very different in the EU in four respects. First, biotechnology products are considered inherently different from conventional products. As a result, the US principle of 'substantial equivalence' is rejected in favour of the 'precautionary principle' to risk management, which emphasizes a cautious approach to new technology when scientific understanding is incomplete. Second, the EU approval process provides much greater scope for the consideration of non-scientific factors, allowing societal, economic, ethical, and environmental factors to influence risk management decisions. Third, consumers in the EU are considered to be entitled to information about how food is produced; hence, labelling is mandatory for all GM products and ingredients. Finally, there are many more veto points in the EU approval process, and this enables small groups of governments to block the approval of any GM crop or foodstuff. This was graphically demonstrated in 1999, when five EU member states suspended new GM product authorizations until a more robust EU regulatory framework was in place. This was eventually introduced in 2001, when new rules were adopted requiring the labelling and traceability of GM foods and the setting of a threshold above which the presence of GM in food and feed must be advertised on packaging (A. Young, 2003; de Jonquieres et al., 2003). These different regulatory philosophies are summarized in Table 2.2.

Table 2.2. Rival approaches to biotechnology regulation

Aspect	US	EU
View of biotech	Substantially equivalent	Inherently different
Approach to risk management	Sound science	Precautionary principle
Consumer information	Only if the product has unusual traits	Yes
Decision-making style	Administrative	Political
Pre-release notification	Field tests—mandatory Pesticides—mandatory Foods—voluntary	Mandatory
Approval required	Field tests—yes Pesticides—yes Foods—no	Yes
Labelling	Only in specific instances	Mandatory

Source: Derived from Young (2003).

Trade experts fear that the GM dispute could prove costly for all concerned: the EU could face damaging counter-sanctions that will harm a wide variety of unrelated goods; the WTO would stand accused of meddling with the democratic wishes of European countries; and, paradoxically, the US could obtain a 'hollow' victory, considering that 'the issues are fundamentally ones of morality and technology—and they must be settled in the courts of consumer opinion' (Victor and Runge, 2003). The earlier dispute may be a harbinger of things to come in the GM dispute because, years after the US won its legal case against the EU, hormone-treated beef is no more able to find a market in Europe than it was before the US victory. In short, the GM dispute illustrates how regulatory models ultimately reflect political values and cultural conventions, all factors that cannot easily be outlawed as 'trade barriers' by WTO dispute settlement panels.

Although the EU and the US are deemed to have equally damaging farm support systems, especially as regards the effects on developing countries, there is one crucial aspect in which they are very different. A UN-sponsored study of trade policy from the standpoint of development (Rodrik, 2001) drew a useful distinction between two radically different styles of 'unilateralism' in the world economy, one that is aimed at protecting differences, the other aimed at reducing them:

When the EU drags its feet on agricultural trade liberalization, it is out of a desire to 'protect' a set of domestic social arrangements that Europeans, through their democratic procedures, have decided are worth maintaining. When, on the other hand, the US threatens trade sanctions against Japan because its retailing practices are perceived to harm American exporters or against South Africa because its patent laws are perceived as too lax, it does so out of a desire to bring these countries' practices into line with its own. A well-designed world trade regime would leave room for the former, but prohibit the latter.

The great merit of this analysis is that it reverses the conventional wisdom (in which development needs are subordinated to the mantra of trade liberalization) by showing how the global governance of trade would look if sustainable development really mattered. As we can see from this analysis, a more sustainable trading regime would be able to accommodate the 'European model' of multifunctional agriculture, but not the 'global breadbasket' model of the US, a kind of agri-imperialism which is intolerant of cultural diversity and inimical to other countries achieving food security through their own efforts.

By trying to design a sustainable agri-food system which meets the multiple needs of economy, society, and the environment, the 'European model' resonates with what developing countries are trying to achieve through their demands for a 'development box', that is, an agri-food strategy that is consistent with their domestic needs. Regulatory regimes, be they national or global, must reflect domestic political values and cultural conventions; otherwise, they will cease to be credible.

Creating a Voice for the Consumer: The Rise of the Food Standards Agency

As stated above, one of the paradoxes of food chain regulation in the twenty-first century is that while rules and regulations are lightened at the production end as a result of liberalization, they are tightened at the consumption end of the chain. The reason for this can be found in the emergence of a new consumer-driven regulatory agenda which, tentative as it is, signals a decisive rupture with traditional agri-food politics. Farm support systems, as we have seen, were largely designed to protect and promote the interests of producers. By the end of the twentieth century, however, these producer-driven systems began to unravel. Although the rising cost of farm support was attracting a growing taxpayer revolt, what really sounded the death knell for the old productivist alliance was a series of health scares that eroded consumer confidence in the food supply chain and the way it was regulated. In the worst-affected countries the crisis involved more than a temporary loss of consumer confidence in a particular product, in some cases amounting to a loss of public trust in politicians, scientific experts, and the regulatory regime itself.

The lengthy series of health scares, which began with food additives in the 1980s, includes botulism, pesticides, Alar, rBST and other veterinary medicines, salmonella, BSE, *E. coli* 0157, GM foods, foot-and-mouth disease, dioxins in animal feed, toxic cooking oil, and, most recently, Sudan 1, a carcinogenic dye used in processed foods. The combined effect of these food scares was to precipitate a series of demands for a more consumer-friendly regulatory regime that does not merely afford better protection but also, and more ambitiously, empowers consumers through the provision of better information, especially as regards food labelling. Here was the making of a new consumer-driven regulatory agenda which was much at odds with the *ancien regime*, or the multilevel food safety policy system guided by the Codex Alimentarius Commission, which sets standards under the WTO for internationally traded food products. The main features of the *ancien regime* have been summarized by Millstone and van Zwanenberg (2002) as follows:

• UK, European, and Codex institutions with responsibility for setting consumer protection standards were also responsible for industrial sponsorship of the food and agriculture industries and for the promotion of trade.

• The regulatory regimes operated under conditions of official secrecy and lacked proper mechanisms of accountability.

• Policy decisions were taken on the advice of small, closed groups of scientific experts, including many drawn from the industries and the firms whose products were regulated.

• Policy decisions were typically misrepresented as based on 'sound science' with almost all the conflicting policy objectives, implicit framings, uncertainties, and residual risks concealed.

• Policy-makers, both public officials and politicians, were able, and keen, to hide behind their expert scientific advisers. Scientific advisers were often expected to take decisions about which risks were acceptable and how they should be managed, even though those decisions required political, rather than purely scientific, judgements.

The fact that the key institutions at all these levels of governance had a dual remit to promote and regulate the industry at the same time means that 'consumer and health interests have been routinely subordinated to the objectives of furthering the commercial interests of farming and the food industry' (Millstone and van Zwanenberg, 2002).

No European country has suffered more damaging food scares than the UK, where each crisis revealed serious shortcomings in the food governance system. As in the US, in the UK the relationship between government and industry was so incestuous that it fuelled perennial complaints that the Ministry of Agriculture, Fisheries and Food (MAFF) had been a willing victim of 'regulatory capture' by the producer interests that it ostensibly sponsored and regulated. Because of these two factors—the succession of health scares and the lack of public trust in the regulatory regime—the UK is an instructive case of a country seeking to create a more consumer-driven regulatory regime based on trust and transparency. While the details may be peculiar to the UK, the challenge is common to all countries.

The lowest point in the history of food regulation in the UK can be dated with some precision: it was 20 March 1996, the day when, after years of repeated assurances that British beef was 'perfectly safe' to eat, the

Conservative government publicly conceded a probable link between BSE-contaminated food and the Creutzfeldt-Jakob Disease (CJD). Such was the significance of this shocking announcement that some policy analysts have gone so far as to say that it 'unleashed the most damaging science-based political crisis that has ever occurred in the UK or the EU' (Millstone and van Zwanenberg, 2002).

From the evidence unearthed by the UK public inquiry into BSE, it is clear that the timely communication of risk to the consumer was not a high priority in the producer-conscious culture of MAFF. On the contrary, by 1987 there was a 'total suppression of information on the subject', given 'the possible effect on exports', all of which was part of a 'culture of secrecy' in MAFF (Phillips, Bridgeman and Ferguson-Smith, 2000). These findings chime with the claims of Edwina Currie, a former Conservative health minister, who said that MAFF officials were not 'the least interested in public health and felt that their task was to look after the farming industry' (Rowell, 2003). Nothing better illustrates the bunker-like mentality at MAFF than the fact that it insisted on keeping secret for eighteen months its knowledge that BSE had emerged in UK herds (Millstone, 2005).

During the early days of the BSE crisis, the UK government's concerns with the economic viability of the food industry, together with its determination to contain public expenditure, conditioned its interpretation and representation of the risks. Whenever regulatory decisions were taken, the impression was always given that the food supply system was completely safe and that the government was motivated solely by public health considerations, when in fact industrial and financial factors were deemed to be equally, if not more, important considerations in the early days of the crisis (Millstone and Zwanenberg, 2001).

Although the Conservative government (1979–97) was in principle inclined to deregulation, in practice it was forced to tighten the public regulatory regime in the 1990s. Its aim here was to catch up with the *private* regulatory system that had evolved rapidly in the UK in the wake of the 1990 Food Safety Act, which sought to promote 'due diligence' in the supply chain as part of a wider system of corporate self-regulation (Flynn, Marsden, and Harrison, 1999; Marsden, Flynn, and Harrison, 2000).

The BSE crisis, along with a plethora of other food scares, propelled the issue of food governance to the top of the political agenda in the 1990s, and the Labour government quickly committed itself to the creation of a new and wholly independent Food Standards Agency (FSA) after it assumed office in 1997. The establishment of the FSA in 2000 was one of two radical reforms in the system of food governance introduced by the first Blair government. The other was the highly symbolic abolition of MAFF in 2001 and its replacement by a new Department for Environment, Food and Rural Affairs

(DEFRA), the title of which had been carefully designed to exclude the traditional farm support function (Barling and Lang, 2003). Radical reform of the food governance system was long overdue; in fact, even after the FSA was established, a major survey of consumer attitudes found that consumers were still 'suspicious and fearful of the food on their plates and do not trust either food manufacturers, retailers or the government to act in their best interests' (Wrong, 2000).

The FSA was one of a number of food standards agencies planned or launched in Europe in the late 1990s. The FSA's main objective, as set out in the Food Safety Act 1999, is 'to protect public health from risks which may arise in connection with the consumption of food (including risks caused by the way in which this is produced or supplied) and otherwise to protect the interests of consumers in relation to food' (Flynn et al., 2004). Charged with the twin tasks of rebuilding public trust in the food supply chain as well as in the food governance system, the FSA sought to create a new consumer-friendly regulatory regime based on three core goals:

- to put the consumer first;
- to be open and accessible;
- to be an independent voice.

In design terms, the FSA is an institutional innovation in three ways. First, it is not an agency in the traditional sense, but a UK-wide government department with offices in England, Northern Ireland, Scotland, and Wales. However, unlike other departments, it is non-ministerial; hence, there is no Minister for Food Safety, but the FSA is accountable to the Westminster Parliament and the devolved assemblies in the Celtic nations of the UK. Second, the members of the Agency's Board are completely independent and were chosen to provide a wide spectrum of expertise across a range of sectors, including public health, consumer groups, food production and processing, catering, communications, and so forth. Perhaps most important of all, though, is the fact that all fourteen members of the Board were appointed after a free and open public competition, a process deemed to be essential if the Agency is to secure public trust. Third, as part of its commitment to being as open and transparent as possible, all meetings are held in public and members of the public have an opportunity to question the Board on any matter in the FSA's remit (Krebs, 2000).

Although the FSA has had its teething problems, it is fair to say that the plaudits outnumbered the criticisms during its first five years of operation, when it was chaired by Sir John Krebs, a distinguished behavioural ecologist. The hitherto critical National Consumer Council, for example, praised the Agency for operating 'with a transparency unheard of in British politics' (Maitland, 2005). In general, the FSA represents a significant improvement on the regulatory regime operated by the MAFF in two major ways. First, it has overcome the regulatory conflict of interest within MAFF, which was linked to its dual responsibility for trade promotion and consumer protection, by becoming unequivocally responsible purely for the latter. Second, the FSA has helped to restore trust in the food governance system by making it more open and transparent than at any time in the past. Sir John Krebs, the first chairman of the FSA, was understandably keen to learn the lessons of the BSE inquiry, the most important of which, he argued, were 'about openness and honesty in dealing with uncertainty and risk' (Krebs, 2000).

Notwithstanding its duty to champion the cause of the consumer, however, the FSA has adopted two controversial policy stances—on the nutritional quality of organic food and on the traceability of GMOs in food and feed—which, according to some critics, have compromised its position. In the Krebs era (2000–5), the FSA seemed to take a perverse delight in antagonizing every interest group in the food universe, from the industrial food lobby, with whom it clashed over salt in processed food, to the green lobby, which was angered by the Agency's statement there was no evidence that organic products were safer or healthier than conventional food products. The fact that the FSA had clashed with the two extreme ends of the food business seemed to confirm, at least to Krebs, that the Agency was genuinely independent, motivated by scientific evidence rather than by powerful or fashionable lobbies. In the case of organic food, however, Krebs was eventually forced to backtrack to a certain degree, by conceding that organics were relatively free of pesticides (Krebs, 2003).

Critics also drew attention to the fact that the FSA's stance on organics contrasted sharply with its more positive stance on GM food. The main evidence for this charge rests on the fact that the FSA opposed the position which both consumer groups and retailers had adopted in favour of the European Commission's proposals to require full traceability of GMOs in food and feed along the food chain. The FSA urged the government to oppose the EC's proposals as unworkable, thus allying itself with the corporate agri-business sector, especially the American Soybean Association, which had the support of the US government. Consumer groups were quick to condemn the Agency for 'having a complete blind spot on GM issues', while other critics felt the Agency was simply falling into line behind its political masters because the Labour government was keen to promote the UK as a world class research location for biotechnology (Barling, 2004).

Since 2003 the FSA has been extending its reach from the more traditional areas of food safety—such as BSE safeguards and chemical contaminants for example—into the more contentious arena of diet and nutrition. Although diet and nutrition were part of the Agency's original remit, their significance has grown exponentially in recent years as a result of the 'moral panic' surrounding obesity, especially the burgeoning problem of childhood obesity (House of Commons, 2004; Morgan, 2004*a*). Sir John Krebs surprised his NGO critics (some of whom had accused him of not doing enough to counter the ill-effects of foods high in fat, sugar, and salt) by taking a very robust stance on the marketing of junk food to children. Among other things, he urged the food industry to recognize that a new, more health-conscious era was upon them, and he called companies that were unable or unwilling to adapt, 'corporate dinosaurs' (Maitland, 2005).

Of all the FSA's activities, the issues of diet and nutrition look set to loom larger and larger in the future, as the government grapples with the spiralling costs of diet-related diseases and as consumers become ever more anxious about the health aspects of their diet. Although the FSA had begun to address these issues in the latter part of the Krebs era, the main problem lay with the government rather than with the Agency, because, as two leading critics argued, 'human health lacks a central position in food policy thinking' (Lang and Rayner, 2003).

One of the biggest challenges facing the FSA in the future will be to develop a more holistic approach to its regulatory responsibilities, not least by forging connections between more quality-conscious consumer demands and more quality-driven producer strategies. The current review of EU Food Labelling legislation offers a major opportunity for the FSA to do so. An important aspect of this review is the issue of *country of origin* labelling. UK farmers, as well as many consumers, are keen to see an extension of mandatory labelling to more foods, processed meat and dairy products in particular. Better country of origin labelling would clearly aid more rapid product recall, but it would also be a statement about a number of other issues, including taste, quality, animal welfare, and environmental integrity. The FSA's own consumer attitudes survey shows that consumers are increasingly looking for information as to where the animal was reared, rather than simply where the food was produced and packaged. The country of origin issue is one on which consumers and producers are beginning to unite: UK consumers want better food labelling information, particularly as regards provenance and traceability, while UK producers want their customers to know that the higher price reflects higher standards of production (NFU, 2004).

Animal welfare and the environment bring us back to the 'non-trade' issues that have bedevilled the negotiations in the Doha world trade talks, highlighting the ever growing links between global and local agri-food agendas. The real significance of the FSA lies in the fact that, for the first time in the history of agri-food politics in the UK, the consumer has a major voice at the highest level of the regulatory regime. Other countries are developing their own food standards agencies, and the EU launched its own European

Food Safety Authority in 2003 (Barling, 2004). These public food govern-ance systems will need to meet two challenges in the years ahead: to prove that they are truly independent of government and to demonstrate that they can keep abreast of the private food governance systems that we examine in Ch. 3.

3

Geographies of Agri-food

Introduction: The Forces of Deterritorialization and Reterritorialization

Chapters 1 and 2 have reviewed the contemporary theoretical and policy context of agri-food with specific reference to Europe and North America. In this chapter we turn our attention to the nature of the new agri-food geographies. What are the driving forces behind these geographies, and how do they play themselves out across time and space? This theme is central to the more detailed treatment of three different regions (Tuscany, California, and Wales) in succeeding chapters. Here, we introduce a conceptual framework that helps us to understand the new agri-food geographies. The chapter starts by outlining the nature of the conventional agri-industrial system. In general terms, we see this as a system that leads to a process of deterritorialization of foods. That is not to say that it comes without any actual geography; rather, its geographies are the result of corporate capitals' attempts to continue to intensify and to appropriate some of the functions of agriculture in ways that stretch the links, networks, and chains between production and consumption spheres.

We then place this trend in conceptual juxtaposition with the more recent forces of reterritorialization (or what some scholars term 'relocalization'), a process whereby local and regional geographies come back again to play a central role in reshaping food production and consumption systems. We argue here that it is important to see these conflicting geographical forces as distinctive, even though both processes may indeed be operating—to varying degrees and in different ways—in the same region or locality at the same time. This is at the heart of our contingent notion of 'worlds of food'.

The New Agrarian Question: The Spatialities of Conventional Agri-industrialism

Throughout the twentieth century, agri-industrialism struggled with resolving Kautsky's formulation of the agrarian question, that is, how to continue to intensify production and appropriate some farming functions in processing and agri-industry while at the same time maintaining some sort of ecological or natural balance in the agricultural transformation process (Kautsky, 1988; Goodman and Watts, 1997). In the agri-industrial model, the driving force was corporate capital. Through an increasing application of science, technology, and capital to food processing, farm input, and farm finance systems, the farm sector was to become increasingly dependent upon upstream (e.g. input-suppliers) and downstream (e.g. food manufacturers and retailers) sectors. Such sectors, in turn, became highly concentrated, while leaving the obstacles of natural and spatial production to a weakened but still (at least in property terms) independent family farming sector.

During the twentieth century this dominant model of agri-industrialism failed to fully appropriate the farm-based sector. However, it conditioned its existence by maintaining increasingly strong arm's-length control of its activities through the operation of a continuous 'cost-price squeeze' and the dynamics of the 'technological treadmill' (Cochrane, 1993), which has cyclically forced independent producers to adopt new economies of scale and labour-saving technologies. This triggered a series of highly specific global–local processes, whereby the tendencies towards 'deterritorialization' and heterogeneity were constantly operating in dialectical fashion in different regional spaces.

The long-run tendency under this 'regime' has been for world agricultural production to grow faster than demand, leading to a decline in international food prices (Mitchell, Ingco, and Duncan, 1997). Early work in the political economy tradition tended to see agri-food following its industrial counterparts down a path of globalization, as defined by the reconfiguration of markets, deterritorialized corporations, and new forms of transnational corporate and inter-firm organization through strategic alliances and networks. In reality, however, agri-food does not follow the same pathways as automobiles and electronics (Bonanno, 1994; Friedland, 1994). In fact, the agri-industrial complex is not simply characterized by vertically integrated transnational production systems, even though many of these tendencies are evident in the workings of such globalized agri-food firms as ConAgra and Cargill (see Goodman and Watts, 1997). Rather, there are increasingly varied forms of corporate international production. For example, even though firms such as Coca-Cola, McDonald's, Kellogg, Nestlé, and Unilever promote global brand names and marketing strategies and rely on transnational forms of integration and corporate control, most of their production is locally and regionally based. This promotes peculiar forms of globalization in the agri-food sector that, as we shall see, set the conditions for variable types of agri-food geographies.

During the 1980s, a rise in neo-liberalism ideology and the simultaneous dismantling of former state and nation-state regulatory authorities (such as Bretton Woods) spurred on a diversity of 'liberalizing' and reregulatory tendencies in the agri-food sector, transforming many national systems of protection and allowing the globalization of agriculture and food to advance in tandem (Buttel, 1997). The demise of national and agrarian forms of Keynesianism has differentially exposed both producers and consumers to the forces of externalization and globalization. In this sense, it is the relative degree of spatial exposure to global forces that has conditioned new forms of agri-food geographies, at least over the past twenty years. This has been perhaps more noticeable in the south of the world than in the north, and in North America (and Australasia) more than in Europe. Nevertheless, Buttel (1997: 346) identifies a generic trend that has gathered speed in all regions, as we outlined in the previous chapter. In his words:

as a result of these national and global shifts, there has been a growing exposure of farm and agri-business enterprises to naked (global) market forces, a return to a more rapid decline in farm numbers, 'industrialisation' of agriculture . . . and the associated restructuring of commodity chains across national borders. . . . Concentration of production is rapidly increasing. For example, of the 1,925 million farms enumerated in the US 1987 Census of Agriculture, 3.6 per cent, the largest of these farms, averaging roughly 28,000 acres, accounted for 50 per cent of US farm output. Only eighteen years earlier (1969 Census of Agriculture), 8.1 per cent of farms, averaging about 1,610 acres, accounted for 50 per cent of national farm output.

It is important to see these globalizing trends in the context of the growth of an ideology of neo-liberal markets and the rise of a concentration in corporate control of agri-food.

Super-concentration and the 'Hourglass': The Integrated Nature of Food Firms

Heffernan, Hendrickson, and Gronski (1999) conceive of the agri-industrial system as an 'hourglass' whereby thousands of farmers feed millions of consumers through an increasingly corporately controlled agri-food system that involves input suppliers, food processors, and retailers. Much of the American literature has focused on the corporate strategies of food firms in the US. Today five major seed companies dominate world-wide: Monsanto, Aventis, DuPont, Syngenta, and Dow. In the US, four firms slaughter 81 per cent of the beef; four firms own 60 per cent of the terminal grain facilities, and three firms export 81 per cent of corn and 65 per cent of soybeans. In addition, four firms have 46 per cent of the total sows in production and four firms slaughter 50 per cent of all American broilers.

In recent years, new alliances and clusters have developed among concentrated firms. For instance, feed manufacturers have joined alliances with processing and biotech firms with food processors (see Figs. 3.1a and 3.1b). Heffernan, Hendrickson, and Gronski (1999) expose the ways in which such

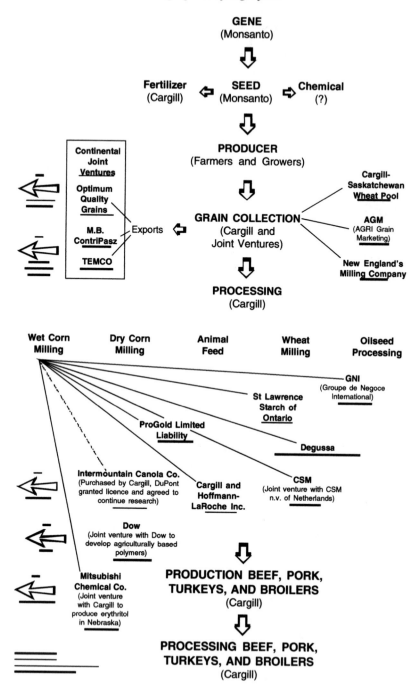

GENE
(Monsanto)

Fertilizer
(Cargill)   **SEED**
(Monsanto)   **Chemical**
(?)

PRODUCER
(Farmers and Growers)

Continental
Joint
Ventures

Optimum
Quality
Grains

M.B.
ContriPasz

TEMCO

Exports

GRAIN COLLECTION
(Cargill and
Joint Ventures)

Cargill-
Saskatchewan
Wheat Pool

AGM
(AGRI Grain
Marketing)

New England's
Milling Company

PROCESSING
(Cargill)

Wet Corn
Milling   Dry Corn
Milling   Animal
Feed   Wheat
Milling   Oilseed
Processing

GNI
(Groupe de Negoce
International)

St Lawrence
Starch of
Ontario

ProGold Limited
Liability

Degussa

Intermountain Canola Co.
(Purchased by Cargill, DuPont
granted licence and agreed to
continue research)   Cargill and
Hoffmann-
LaRoche Inc.   CSM
(Joint venture with CSM
n.v. of Netherlands)

Dow
(Joint venture with Dow to
develop agriculturally based
polymers)

Mitsubishi
Chemical Co.
(Joint venture
with Cargill to
produce erythritol
in Nebraska)   **PRODUCTION BEEF, PORK,
TURKEYS, AND BROILERS**
(Cargill)

**PROCESSING BEEF, PORK,
TURKEYS, AND BROILERS**
(Cargill)

Fig. 3.1a. Cargill/Monsanto joint ventures and strategic alliances
Source: Heffernan and Hendrickson (1999).

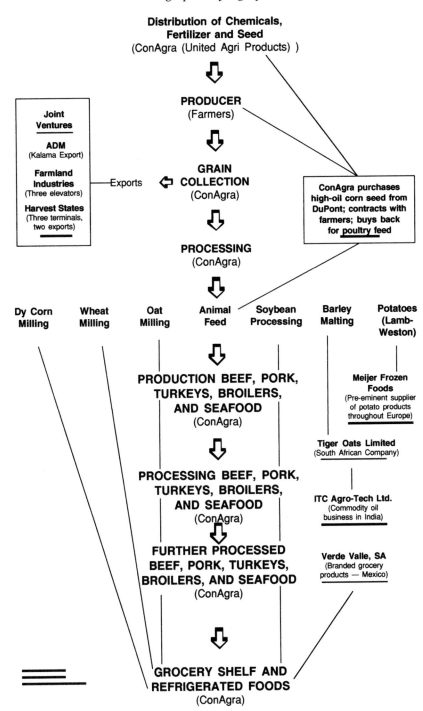

Fig. 3.1b. ConAgra joint ventures and strategic alliances
Source: Heffernan and Hendrickson (1999).

alliances obscure and temper market mechanisms. They explain that there is little 'price discovery' during the transitions between the gene, fertilizer, and processing to the supermarket shelf and that 'the only time the public will ever know the 'price' of animal protein is when it arrives in the meat case' (p. 3). Under these conditions, the farmer becomes a 'grower', being autonomous only by dint of his/her landholding and of the need to cope with any vagaries of nature that may get in the way of an otherwise seamless process of production and consumption. The food product is passed from stage to stage, but the location of control and decision-making remain focused in one place.

One cluster that brings different rural and urban spaces together in a complex integrated chain extending from the gene to the grower and the processed broiler (see Fig. 3.1a) is the *Cargill/Monsanto cluster*, which is the result of a joint venture started in 1998. While operating seed and research operations in twenty-three other countries, Cargill did not have access to gene technology. Hence, it formed a joint venture with Monsanto, a company that had the intellectual property rights to develop genes. For Monsanto this implied that it did not need to rely upon access to farmers' fields to test products (such as the 'Terminator gene'). There was thus a mutual need in the development and implementation of technologies to extend control over the entire food chain.

The processes of firm acquisition and 'clustering' are highly dynamic. In the late 1990s, the acquisition of 'Continental grain' allowed Cargill to gain control of 40 per cent of all corn exports, a third of all soybean exports, and at least 20 per cent of wheat exports. As a result, in certain regions such as Illinois, Ohio, and the Mississippi plains, farmers' options for selling their grains are severely limited. Cargill's corporate goal is to double its size every five to seven years, with the expectation that there will be a 20 per cent return on the equity tied up in the whole cluster (Heffernan, Hendrickson, and Gronski, 1999).

A second cluster centres on *ConAgra*, a firm projecting the image of dealing with the whole process from 'the farm gate to the dinner plate' (see Fig. 3.1b). *ConAgra* is now one of the three largest flour millers and corn millers in North America, ranking third in cattle feeding and second in cattle slaughtering, third in pork processing and fifth in broiler production and processing. In 1998, it formed a joint venture with ADM and DuPont, linking itself to seed production and exporting bodies. ConAgra now sells branded food products as well, and it is the second leading food processor in the US. The growth of ConAgra during the 1990s was built upon a strategy of acquisition, divestment, and value-adding product lines.

The development of these integrated clusters, especially in the US, is based upon a new imperative to bring together biotechnological facilities, grain and feed trading, and food meat processing. In this context, the more traditional (and linear) commodity-systems approach to agri-food is now somewhat dated (Friedland, 2001); in fact, the relationships between commodity sectors and their functions have become as important as the particular frameworks surrounding the specific commodities themselves. In addition, these integrated clusters allow for 'price-hiding' markets in the transfer of materials which go to make up final food products. This further compromises the situation of individual producers, as it weakens their bargaining positions and forces them to sell to a handful of buyers. For most livestock commodities in the US, the production stage is now highly integrated into the processing and the larger clusters described. For instance, 95 per cent of broilers are produced under contracts with fewer than forty firms. Twenty feedlots (processors) feed about half the cattle in the US and are owned by the slaughtering firms themselves or by the producers who have contracts with the processing firms. While the production sector stubbornly remains outside these integrated agri-food clusters, it is increasingly dependent upon them for its inputs and outputs. The 'cost-price squeeze' that farmers face is now increasingly surrounded by the operation of 'non-markets', that is, places where transactions occur but where there is no legitimate 'price-discovery'. Meanwhile, concentration continues apace in the major agri-food commodities. Heffernan, Hendrickson, and Gronski (1999) have demonstrated that this level of concentration has significantly increased throughout the 1990s in beef packing, feedlots, pork packers, etc.

While on the consumption side of the equation these tendencies lead to essentially 'delocalized', standardized, but more differentiated product ranges on supermarket shelves, it is important to consider that they have also created more homogenized regional and local geographies of production based upon fewer but larger holdings.

Agrarian Regionalization

As FitzSimmons (1997) argues, the restructuring of conventional systems of agri-food requires a renegotiation of spatial relationships at multiple geographical scales. In the American literature, it is often argued that the power to recreate these geographies has been centred upon the dominance of agribusiness and its influence in shaping regional agri-food geographies. Fitz-Simmons (1997: 156) states that

connections between input suppliers, farmers, and processors and marketing firms are likely to reflect multiple spatial relationships at several scales and may include regional monopolies or oligopolies in input and product markets. While global capital and risk markets have now appeared, most actors in farming systems are connected to these markets through a chain of intermediaries; only the largest transnational firms access these markets directly. Government policies may defend internal social practices and established alliances or, in the new era of liberalization of agricultural trade, encourage export production of some farm commodities at the cost of relaxing trade barriers that defend other agricultural products.

The geographies of concentration and integration in the agri-business sector have thus created a highly uneven pattern of regional specialization in production. Geographical specialization, monoculture, and spatial homogeneity have been the key forces in both North America and in Europe. This has created platforms of farm production that can provide standardized products at a large and constant scale according to the specific standards required by manufacturers and retailers.

As a consequence of this situation, both in the US and in Europe there has been a continuous decline in mixed farming systems (Buttel, in press). In 1900, in the US over 75 per cent of farms raised cattle and milked cows. By 1997, only 6 per cent of all farms had milk cows and only 6 per cent raised pigs. We shall examine some of these trends with regard to specific regions in succeeding chapters; here, however, it is important to provide some general insights on this continuing regional specialization, whereby horizontally, crops and livestock have become increasingly separated and the 'efficient' production of chickens, dairy goods, and potatoes has been restricted to the largest and most capital-intensive holdings; vertically, on the other hand, there is increasing arm's-length control by agri-business of the farm-based sector. In what follows we examine these trends with regard to three case studies provided in the literature.

Pork Production in Iowa

A general trend has been to develop or, more accurately, to superimpose an integrated and monocultural tendency on regional specializations that had occurred historically, largely because of natural, geographical, and market advantages. One example of this is hog production in Iowa (Page, 1997). The growth in the oligopolistic power of agri-business is leading to the continued transformation of the region's principal agricultural product in four ways. First, after large-scale southern contracting firms entered the state, despite a strong ideology of family farm independence, more hog farmers found it necessary to grow to contract. This reduced their independence and forced them to meet the standards of the food processors. Second, the relationships between farmers and input suppliers changed, with some local feed firms integrating forwards into contracting and production. Moreover, many larger contracting firms are privately and internally managing their own research and development, bringing into question the role of the traditional links between the land-grant universities and the farming population. Third, there has been a political opposition to the growth of corporate hog production, and this is influencing the location of the hog-pork commodity chain. As a result, some firms are moving just outside the boundaries of the state. Fourth, firms such as ConAgra, Cargill, Smithfield, and IBP (the nation's premier pork-packing firms) have large investments in Iowa hog slaughtering. Through their power, they have ensured that their farmer suppliers produce more consistent lean and uniform hogs, and this is increasing the size of farms. As a consequence of these transforming tendencies, pork production in Iowa remains 'extremely unsettled' (Page, 1997).

Meanwhile, between 1989 and 2002 the top four pork-processing firms increased their market share in the US from 34 to 59 per cent. Hendrickson and Heffernan (2002) state that Smithfield, for instance, 'followed the broiler model with significant acquisitions in both processing and sow production'. Now this firm is looking overseas, and the Iowa model is replicated by production facilities in Mexico, Poland, Canada, and Brazil through joint alliances with Artal Holland BV.

The Broiler Filière in the American South

The broiler industry on both sides of the Atlantic is of considerable dynamism and growth and it continues to reshape regional agri-food geographies. Boyd and Watts (1997) analyse the evolution of this filière from the 1930s to the end of the century and show that innovations in breeding control, disease management, nutrition, housing, and processing have been the driving forces for lower production costs. Breeding, in particular, reduced the need to feed the birds, while also increasing the average live weight from 2.89 to 4.63 pounds and decreasing the maturation period—the period for the bird to reach 'market weight'—from 70 days to less than 50 (G. Watts and Kennett, 1995).

While the patterns of inter-farm dependence between the 'integrators' and the complex field of suppliers-breeders, processing equipment, etc. resemble those depicted in the wider industrial geography literature, with more flexible marketing patterns and just-in-time systems of delivery to outlets (see Amin, 1994), Boyd and Watts clearly distinguish these patterns from those associated with 'Toyotaism', given the inevitable 'organic' nature of the product and the severe working conditions of many of the employees. Moreover, they point out that just-in-time production has not resolved the endemic problem of overproduction, which periodically plagues intensive industries. They argue (1997: 215) that 'systematic overproduction is rooted in the fact that the broiler industry, like many industries dealing in food and natural resource commodities, operates on the basis of very low margins. The reality of low margins means that profitability rests on 'turning volume' by expanding capacity and increasing productivity, which itself requires the successful venting of surplus production'. This logic has forced leading firms such as Don Tyson to develop a policy of 'segmentation, concentration and domination' and to further diversify their product ranges on the market as a way of 'venting surplus'. From its original and traditional base in the southern US, a global industry involving Brazil, Thailand, and western Europe has now developed (Flynn, Marsden, and Smith, 2003). The original rise of the southern production complex was built upon a radical restructuring of the entire complex that can be traced as far back as the 1960s, when a critical 'just-in-time' system of industrial organization became regionally based. Central to this system was the 'integrator complex', which brought together hatcheries, breeder grows, feed mills, broiler grow-out, rendering, processing, and further processing. Tyson's has 27 such integrated broiler complexes in the South, which include 45 hatcheries, 28 feed mills, 55 processing plants, 7 distribution centres, 6 rendering plants, and 15,800 'grow-out houses' with a capacity of 35 mill placements in one week (Thornton, 1996).

Intensive Production in the Netherlands

Van der Ploeg (2003) has analysed the historical process of upscaling in the Netherlands, where the bulk milk dairying sector has been encouraged by the Ministry of Agriculture to develop economies of scale under a 'free-trade scenario' that predicts fewer and fewer farmers. Van der Ploeg attributes the dominance of this growth paradigm to an amalgam of interests, involving the Dutch Ministry, research and development sectors, and the agri-business complex (such as Friesland dairy foods), which promotes scale, volume, and standardization, now based upon strict hygienic criteria. The upshot of this amalgam of interests is the notion that a viable farm in the region in Friesland is capable of producing 800,000 litres of milk per year. This was identified as the only objective for the Dutch dairy sector in the context of GATT and WTO on the one hand, and as a consequence of the enlargement of the EU on the other. It was argued that about 7,000–8,000 greatly expanded dairy farms would be able to provide 60–70 per cent of the total Dutch milk quota.

The process of upscaling has accelerated most sharply in intensive husbandry and greenhouse horticulture. In addition to promoting regional homogenization, van der Ploeg (2003) argues, this process was also an important social source of contestation and conflict, with many independent farmers trying out different farming styles in the same region and the local and regional state bodies attempting spatially to compartmentalize the industrial forms of agriculture through land-use planning. In other words, this process led to a contested agri-food space in which the guiding logics of upscaling, technological development, and integrated non-farm processing and input suppliers play a key role.

In short, these examples depict a wider process of radical reorganization and integration of the conventional agri-food sector in the US and in many parts of Europe. Although this has affected particular agricultural commodities differently, the general trend has been to create agriculturally intensive super-regions based upon integrated systems of production and processing. This has furthered the demise of both mixed farming systems and regionally based agri-ecological systems (van der Ploeg, in press). These trends are also based upon the long-running 'Kautskian compromise' in agrarian capitalism, which enables farm holdings to be left as quasi-independent enterprises that are increasingly dependent upon upstream and downstream corporate sectors.

The phenomenon of super-intensive agri-regions is built upon ever growing thresholds of scale and intensity, which, from the medium-term time horizons of the farmer, represent a sort of competitive 'race-to-the-bottom' with regard to the balance between costs and farm gate prices. These are then regions and geographies which eventually devalorize production by creating overproduction and devalorize nature by continuing to 'manage' its vagaries through the continual application of technological 'fixes'.

The Power Shift: From Food Manufacturing to Corporate Retailing

The agri-industrialization explored so far and its profound consequences for spaces of production and consumption have been experienced on both sides of the Atlantic. Even though the levels of inter-firm integration in Europe have been hindered by the boundaries of European nation-states, after the establishment of the European internal market over the past decade, more cross-border integration, as well as deeper food trade, have been a hallmark. In fact, the food and drink sector in the EU is the largest manufacturing sector, accounting for 13 per cent of the total production, with France, Germany, Italy, the UK, and Spain the leading producers of foods and drinks in the EU, representing 80 per cent of the total production.

Despite the growth of global corporations (such as Unilever), the EU sector is characterized by a large number of small and medium-sized enterprises (SMEs). In 2000, these represented 99.3 per cent of the companies in the food sector and generated almost half its total production and 62 per cent of its employment. These trends are more pronounced in the southern European countries; in Italy, for example, four out of five food employees work for an SME (CIAA, 2003).

The contemporary growth in food consumer markets, the increasingly weakened and fragmented position of the food and drink manufacturing sector and the spectacular rise of corporate retail ('merchant' capital, rather than industrial capital) have been the hallmarks of the European agri-food sector. Partly as a result of these trends, food manufacturers have found difficulty in maintaining and promoting their branded products *vis-à-vis* their competitors in the US (Cotterill, 1997). Consequently, and despite similar processes of consolidation and vertical integration among the larger food manufacturers, the more consumer-based corporate retailing sector has been the driving force in reshaping the European agri-food sector. As a telling review of the UK food manufacturing sector recently reported (Fenn, 2004: 41):

the UK food industry is dominated by very large (often multi-national) companies, which operate across a range of food markets and are often vertically integrated. The consolidation process continued in 2003 and 2004, exemplified by the takeover of Weetabix, the last major UK-owned cereal manufacturer, by the US venture capitalists, Hicks, Muse, Tate and Furst (which already owned a major stake in Premier Foods).

Consolidation is driven by the intensely competitive nature of grocery retailing in the UK. Mergers and acquisitions, which can boost economies of scale, are one way of lowering costs, but the increasingly cut-throat nature of grocery retailing inevitably results in some casualties. In October 2003, for example, Hibernia Foods, the owner of iconic brands such as Sara Lee, Entenmann's, and Mr Brain's Faggots, went into receivership, blaming excessive discounting by the supermarket chains.

Several writers have compared the corporate power of agri-business in the US with that in Europe (Wrigley, 2002; Wrigley and Lowe, 2002) and have focused on the vibrancy and increasing corporate power of European retailers. The main question here is: how does the uneven growth in retail corporate capital affect the recomposition of spatial relationships with regard to agri-food? We can delineate four significant trends recently identified in the literature which, we believe, require further cross-national and comparative research. These include:

- the development of consumerized corporate retailing;
- the concentration and market power of the retail sector;
- retailers' ability to manage space–time relations;
- geographies of quality control.

The Development of Consumerized Corporate Retailing

For almost fifty years (from the 1930s to the 1980s), regulation in the US was largely hostile to the development and the concentration of corporate retailing; as a result, very little change occurred during this period with regard to the market (Wrigley, 2003; Wrigley and Lowe, 2002), almost no power was shifted to the retailers from the food manufacturers (Hughes, 1996) and, as we shall see, the retailers largely remained regionally focused. This scenario has now begun to change significantly, and it is doing so in a context where the large multinational food manufacturers are maintaining strong brand and spatial presence.

The development of a vibrant merchant retail capital, and its imprimatur upon the geography of agri-food, has been uneven in the US compared with Europe. Nevertheless, the last decade has seen a growth in the overall internationalization of retailers (Busch, 2004). Wal-Mart first opened a shop outside the US (in Mexico City) in 1991; Tesco had its first hypermarkets in 1997; and Carrefour started international business in 1989 in

Taiwan. Today, Wal-Mart has 5,070 stores in ten nations, Carrefour has 10,378 in twenty-nine countries, and the Dutch Royal Ahold has 5,006 stores in Europe, North America, Asia, and Latin America (Hanf and Hanf, 2004).

All retailers now operate on several continents both in terms of locating stores and sourcing products. Hence the agri-food system is experiencing a second wave of globalization—one that Kautsky, writing at the end of the nineteenth century, was unable to imagine and that scholars have tended to underestimate (see Bonanno et al., 1994). Whereas the literature on globalization of agri-food was represented as essentially part of just the food manufacturing sector, today a dynamic of internationalization of corporate retailers is superimposing itself on this picture.

To understand the emerging agri-food geography, it is therefore important to recognize some of the key and distinctive features of retail capitalism (Marsden, Flynn, and Harrison, 2000). Unlike food manufacturing, this is not based upon the transformation of an agricultural product, but on its sale to a widening group of consumers. Moreover, unlike other branches of agri-food, retail capitalism is less dependent upon the natural and organic constraints of agriculture and food—even though it is still affected by the perishability of food products and the material and financial 'waste' this creates. In addition, and crucially in an advanced world where overproduction is the norm but the proportion of household income spent on food goods tends to decline, retail capital becomes pivotal and sits strategically between the consumer and the supply side. These distinctive features mean that the *raison d'être* of retail capitalism becomes the effective and increasingly complex procurement and trading of foods for different segments of consumers. As a result, retailers spend much of their time (as an economic and logistic imperative) both constructing and reflecting the 'consumer interest', leaving the still problematic and 'awkward' organic and natural transformation of products to the upstream producers and the manufacturers.

Allaire (2004) considers this retail-led process as a distinct and mutating 'cognitive paradigm', operating around relative 'quality' conventions that have significant social impacts. This is, for him, a decomposing and recombinative process of knowledge creation, innovation, and transfer. He argues (p. 85):

as a pattern of complementary innovation, I see decomposition/recombination applying to agronomic inputs and food-end products by entailing a conception of an agriculture and food premised on a splitting down of information and material into discrete 'bits' in a serial linear process. The decomposition paradigm gives valuable representations to the evolution of information structures in the agriculture and global food system of innovation premised on the collection of highly detailed data through continuous automated monitoring at all stages of production and marketing. According to this paradigm the creation of new products, markets and consumption arenas is predicted by permutations within rapidly expanding databases. Communication technologies in the production and marketing domains, such as bar-coding, electronic data interchange and precision farming, contribute to overall database integration. These technologies offer a potential means by which to enhance efficiency and the capacity to differentiate over time through a kind of automated learning.

In summary, it is clear that, in Europe especially, particular regulatory, technological, and market conditions over the past twenty years have been highly conducive both to the rise in the scale and geographical scope of corporate retailing and to the increasing concentration and power of the retailers *vis-à-vis* food manufacturers and producers. Partly as a consequence, supply chains have become more geographically stretched and knowledge-based.

The Concentration and Market Power of the Retailing Sector

Supermarket groups now account for almost half of all food retail sales in Europe. Between 1999 and 2003, the sales of all goods by food retailers in nineteen European countries climbed rapidly from 16 to 46 per cent of the total retail market (Fenn, 2004). In the UK, Tesco sales grew by 54 per cent between 1999 and 2003, due to a 65 per cent increase in floor space. After taking over German and British firms, Wal-Mart increased its sales by 32 per cent over the same period. Carrefour, the French-owned retailer, remains Europe's largest retailer. In the US, as we indicated above, even though this trend started later, consolidation is now well under way (Wrigley, 2003). Since the late 1990s, five retail firms (Kroger, Albertsons, Wal-Mart, Safeway, and Ahold USA) account for 42 per cent of retail sales, whereas in 1997 the figure was 24 per cent. Through acquisition and reconsolidation, the traditional regional structure of supermarkets in the US is giving way to national and, increasingly, international players.

There is now a growing literature on the internationalization and concentration of the corporate retailing sector. It is clear that these trends will continue to restructure the food supply system not least by reinforcing the oligopolistic buying and selling power of the retailers *vis-à-vis* the food manufacturers and producers. This is now affecting the processes of cluster development and integration in the agri-business sector discussed above. For instance, in the US, Kroger has developed a case-ready beef-supply agreement with Excel (Cargill), while Stop and Shop (Ahold USA) has a dairy supply agreement with Suiza foods and Wal-Mart obtains case-ready meats from IBP, Farmland, and Smithfield (Hendrickson and Heffernan, 2002). Increasingly, even the largest food manufacturers realize that they need to establish relationships with international retailers and to monitor their actions very closely.

Retailers' Ability to Manage Space–Time Relations

As a cause, and partly as a consequence, of these recent trends, there has been a development of consumer-based supply-chain management systems which enable the retailers to manage space and time. Most obviously this occurs with the development of own-brands, which directly compete with the more established brand names of the large food manufacturing conglomerates. Busch (2004) has indicated that this has stimulated the growth of a new class of food processors who specialize in producing to supermarket specifications. The rise of own-brands, pioneered in the British retail sector, represents over 50 per cent of all supermarket food goods sold in some retail outlets in the UK (Marsden, Flynn, and Harrison, 2000), while in the US and Canada the proportion is reaching 25 per cent. This is a direct challenge to the established food manufacturers, and 'brand warfare' is now a key feature of the agri-food system.

Another key trend in the retail sector is to stock and procure more fresh foods, thereby further reducing the need for branded goods. The rise of the global fresh food and vegetable sector has been largely a result of this retail innovation in supply chain management, which has enabled the corporate retail sector to circumvent or short-circuit the traditional manufacturers and to develop sophisticated supply chain arrangements with specific regional production 'platforms' (Busch, 2004; Marsden, 1997; Raynolds, 1997).

These retailer-led developments have profound implications for the management and recomposition of space and time from the farm to the home. A key priority is 'just-in-time' delivery to the supermarket shelf; these time logistics need to be balanced with the ability to manage sales so as to shift the product to the customer as fast as possible. Hence, 'shelf life' and 'category management' become critical elements to minimize waste in the supply system. A major competitive edge over rivals can be gained by innovations in the management of space–time. In the UK, Tesco's success and Sainsbury's current crisis have been partly associated with the former's ability to apply effective space–time logistics. This new 'science' is now a significant part of many leading business schools as it constitutes a new business model.

Geographies of Quality Control

These new space–time rules and conventions reshape certain regional production spaces and include them within their ambit. This becomes a system of arm's-length control of supply through new sets of intermediaries (such as certification bodies, dealers, transport firms, distributors) that place stricter control on the particular quality of the product. This is clearly seen in the fresh fruit and vegetable sector (Cavalcanti and Marsden, 2005; Marsden, 1997; van der Grijp, Marsden, and Cavalcanti, forthcoming), where corporate retailers have developed collaboratively their own quality protocols (such as EUREPGAP) to be enforced at the point of production. A condition for market entry, for instance, for Brazilian mango and grape producers is the requirement to meet standardized and increasingly demanding protocols, which can be considered new forms of non-market private regulation and governance. Such protocols are also a source of major dispute at the WTO (see Ch. 2), with developing countries arguing that they constitute new types of non-tariff barriers in a supposedly more liberalized food market.

As Cavalcanti and Marsden (2005) and van der Grijp, Marsden, and Cavalcanti (forthcoming) point out, this is an increasingly exclusionary process. In fact, only the larger and more expert producer and export firms are able to meet these increasingly demanding criteria. In this sense, as we argued with regard to the large agri-business clusters, we see the attenuation of transparent pricing, spot markets, and traditional food brokerage. Competition is based upon meeting retailer-led conventions. Moreover, in an increasingly internationalized sector, where retailers are capable of sourcing globally but selecting spatially, such quality standards and protocols allow them to enforce these systems in different locations at the same time. Hence, a recent innovation has been introduced by the retailers in the sphere of what they call 'non-competitive' quality protocols, which allow them spatially to choose production locations, while also ostensibly maintaining quality standards in their stores. This can hold severe implications for those (often neighbouring) farmers who lie outside these quality controls and it creates a new dualism in the countryside.

Hendrickson and Heffernan (2002) have explored these new retailer-led space–time relationships with regard to the US dairy sector. They argue that retailers are now in a position to dictate terms to food manufacturers, who then force changes back to the farm level. Between 50 and 75 per cent of the total net profit for large retailers comes from fees from 'slotting allowances', 'display', 'presentation', 'pay to stay', and 'failure'. Both in the UK and in the US (Marsden, 2004a), the dairy sector has been particularly vulnerable to the imposition of this retailer-led system. In the UK, the average price differences between farm gate and retail prices have grown significantly since the BSE crisis, as we will see in Ch. 6. It is estimated that farmers receive 26 per cent of the retail price for beef, 20 per cent for pork, 21 per cent for chicken, 25 per cent for milk, and only 8 per cent for potatoes (W. Young, 2004). Regional and national cooperatives have been dismantled or absorbed as significantly weaker members of agri-food clusters, at the same time as integration and concentration have occurred in the milk processing sector. As Hendrickson and Heffernan (2002: 2) point out:

vertical integration, which formally connects the dairy processing stage to the retail stage, is probably the major driver of the [dairy] restructuring at this time. Through acquisitions, Kroger and Safeway own and operate their own dairy processing facilities to supply some of their needs. But Kroger and Safeway, as well as Walmart and others, are seeking long-term agreements which guarantee them consistent product to serve a coast-to-coast operation. Most processors now see the retail firms as their consumer.... The three largest global food processors (Nestle, Unilever and Philip Morris) are directly involved in dairy processing in the United States. These firms will 'source' their raw milk from wherever they can obtain it at least cost. Already there are calculations estimating the world price of raw dairy products if the United States produced none. In the past couple of years, the importation of milk protein concentrates, which do not face any import restrictions, has greatly increased.

Hence, while on the one hand the retail-led system drives product differentiation and value-added products based upon specific 'quality' conventions, on the other hand, it sources on the lowest price and continues to empower itself through negotiated quality/time/price arrangements.

Looking into the Eyes of the Hog: Some Conclusions on the Conventional System

What lessons can we draw from these new agri-industrial and corporate retailing trends and the ways in which scholars have depicted them? It is clear, as we have tried to show, that there are significant and profound forces at work. Uneven as they are, they impinge on regional spaces in complex ways and condition their worlds of food. They also, as we shall see below in considering the alternatives, partly shape the competitive spaces on which alternative food networks have to play, in both the North American and European regulatory contexts. There are several points that it is useful to distil with regard to the implications for the new agri-food geographies.

- The conventional system is not static. FitzSimmons (1997) reminds us of the complex layering and multiple institutional complexity of any given agrarian region or space. A key common feature seems to be the continued pressure to devalorize or depreciate rural space through a process of scale enlargement and delivery. This is no longer exclusively associated with the constant appropriation tendencies of agri-business capital, in the form of corporate input and processing firms. Such a feature now finds a new 'level' of appropriation through increasingly globalized mercantilist retail capital (i.e. shifting and trading goods as well as producing and transforming them).

- There are new, highly managed, and logistically complex space–time vectors that redefine the quality/scale parameters of the products.

- The 'conventional system'—itself an amalgam of competing clusters of integrated agri-business and corporate and internationalized retail firms—is now capable of producing a vast array of diversified products (over 18,000 in Tesco, for instance), at the same time as rigidly controlling its globalized supplier networks through 'arm's-length' protocols.

- A key feature and 'triumph' of such global capitalistic coordination is its ability to partition and distantiate itself from the natural and organic geographies for which it is responsible. 'Fresh' fruits and vegetables are placed on shelves, but need to be moved as quickly as they arrive.

Consumers are asked no questions and are given minimum knowledge of their provenance. Distantiation at the point of choice and purchase is as important as the constructed distance created in the quality control frameworks. In short, markets are hidden.

- In this context, it is difficult to see anything other than a disempowerment of the producers. At best, these systems lead to new forms of uneven development between those participating in the global and regional networks and those who are not. At the same time, we see the gradual erosion of former producer-based systems of collective institutionalization in the form of cooperatives and collaborative arrangements. Not only do the economic transformations seem to demand scale, production, quality, and price, they also seem to carry the message of a progressive individualization of producer behaviour.

It would be wrong to suggest, as we have pointed out, that the 'conventional system' of food supply is anything but homogenous or standardized. What we have tried to depict here are some of its key contemporary dynamics: concentration, oligopoly, technologism, consumerism and scale, quality, and spatially distantiated management of supply chains. These innovations provide a high variety of 'choice' for consumers, under conditions where markets are supposedly more 'liberalized'. In reality, there are few transparent markets, and entry to these is increasingly controlled by 'non-market' quality criteria. At the same time, while these systems bring significant benefits to some agricultural regions, they continue to devalorize the primary production sector. In short, culture still drains away from agriculture at an 'alarming rate' (Kropotkin, 1906; Pretty, 2002).

In sum, we are left with a potential regional 'draining' process with regard to the territorialization in the conventional sector. Yet these trends can be resisted, and, as we shall see below, their very existence provides a source of positive struggle to adopt alternative strategies that are designed to rearrange the place, provenance, and power relations associated with agri-food. This raises some fundamental questions: how does such a system continue to survive/flourish and be publically legitimated? What is the significance of alternative food networks and local and regional countermovements? If successful, how can alternative food networks sustain their 'alternativeness' with regard to recapturing value back to the local/regional agricultural domain?

Reterritorialization and Localization and the Geographies of Alternatives: The New 'Battlefield'

It is clear that a key feature of the conventional systems relies upon the definition and implementation of 'quality' criteria. As M. Harvey, McMeekin, and

Warde (2004) and Allaire and Wolf (2004) point out, quality criteria become part of the new heterodox economics and polity of contemporary food supply chains and their regional development. Allaire and Wolf (2004) identify two main paradigms, locked in conflict, which produce many 'hybrids' with regard to quality 'values'. One paradigm, as outlined above, follows the logic of decomposition, a transforming 'Taylorization' of food production and consumption (see also Friedland, 1994; 1997), where each input into production and consumption is seen as an object of innovation. The other flows from a more holistic logic of identity, embracing different dimensions of food aesthetics, ethics, sociality, purity, naturalness, and potential social countermovement.

In abstract terms, we can conceptualize these 'alternatives' as a heterogeneous mix of relocalization of alternative agri-food (see Fig. 3.2). Such alternatives do not by any means have a monopoly on 'quality' food (M. Harvey, McMeekin, and Warde, 2004); in terms of specific food products, they can create a range of hybrids, where different types of quality conventions (commercial, ecological, civic, to use Storper's terms) become blurred. However, we argue here that, from a sociological and geographical perspective, these alternatives are structurally distinct from the conventional agri-food systems described above. As we show in Fig. 3.2, there are some ideal typical features of relocalization and revalorization.

Conceptualizing 'quality' as an intensely competitive economic and spatial 'battlefield' helps us to understand the broader aspects of the contemporary political economy and power relations in agri-food. In this sense, ideas of quality need to be seen in the context of the competitive development and regulation of food supply chains. It is clear that contestations, not least between the two rival paradigms identified in Fig. 3.2, have come to play a key role in preserving and reallocating power relations within particular types of food supply chain. A new literature demonstrates the highly uneven emergence of 'alternative' food supply networks (D. C. H. Watts, Ilbery, and Maye, 2005), which are developing in the interstices of the more conventional retailer-led supply chains, and partly define their actual 'alternativeness' through their competitive relationships with the more conventional system. We represent this dichotomy as a battlefield of knowledge (Allaire, 2004), authority, power, and regulation (Marsden 2004*a*) and, as we will explore in the rest of this chapter, space and spatial competitiveness. That is, there are different and increasingly fluid 'worlds' of food within the same spaces operating rival paradigms of knowledge, power, and regulation (Fig. 3.2).

These rival paradigms fight around distinct social and technical quality conventions. The outcome of this process is to empower or disempower particular sets of supply chain actors. In much of the alternative food networks literature, the key objective for producers is to regain power in the chain and to revalorize primary production against the 'race to the bottom' features of the conventional systems (Renting, Marsden, and

Type of spatial relationships	DELOCALIZATION Conventional agri-food		RELOCALIZATION Alternative agri-food
Producer relations	Intensive production 'lock-in'; declining farm prices and bulk input suppliers to corporate processors/ retailers.		Emphasis on 'quality'; producers finding strategies to capture value-added; new producer associations; new socio-technical spatial niches developing.
Consumer relations	Absence of spatial reference of product; no encouragement to understand food origin; space-less products.		Variable consumer knowledge of place, production, product, and the spatial conditions of production; from face-to-face to at-a-distance purchasing.
Processing and retailing	Traceable but privately regulated systems of processing and retailing; not transparent; standardized v. spatialized products.	CHANGING COMPETITIVE SPATIAL BOUNDARIES	Local/regional processing and retailing outlets; highly variable, traceable, and transparent; spatially referenced and designed qualities.
Institutional frameworks	Highly bureaucratized public and private regulation; hygienic model reinforcing standardization; national CAP support (Pillar I).		Regional development and local authority facilitation in new network and infrastructure building; local and regional CAP support (Pillar II).
Associational frameworks	Highly technocratic—at a distance—relationships; commercial/aspatial relationships; lack of trust or local knowledge.		Relational, trust-based, local, and regionally grounded; network rather than linear-based; competitive but sometimes collaborative.

Fig. 3.2. Rural space as competitive space and the 'battleground' between the conventional and alternative agri-food sectors

Banks, 2003). As a result, one important theorization of food quality becomes associated with the ways in which different supply chain actors compete for the authority and legitimacy of defining its actual character. This is a highly competitive and contested process; one that is shaping not only consumer decisions, but the competitive 'spaces', boundaries, and markets themselves in which both established conventional players and 'alternative' food actors are situated.

Table 3.1 highlights the relationships between the spatial and the quality dimensions and shows the 'quality battles' occurring between the more highly intermediated and extended quality supply chains (which now also involve corporate retailers) and the local face-to-face, or proximate, regional and ecological chains. Such a multidimensional matrix reveals two important tensions. First, regional-artisanal or ecological-natural quality product defi-

Table 3.1. Theorizing food quality—opening up the quality food spectrum: the short food supply chain (SFSC) battleground

	Quality parameters	
SFSCs scale	Regional-artisanal paramount	Ecological-natural paramount
Face-to-face (direct producers–consumers)	Typical products (e.g. speciality cheeses) On-farm processing Farm shops Farm producer	Organic box schemes Farmers' markets Organics
Proximate (some intermediation)	Farm-cottage foods Regional labels Wine routes Special events Local cuisine restaurants New cooperative marketing arrangements	Free-range GMO-free
Extended (high intermediation)	Designation of signs (CPDO-PGI) Fair trade products Ethical products Regional brands in supermarkets	Retailer organics (+70 % in UK) Integrated pest management systems Free-range GMO-free 'Slow food products'

Source: Marsden (2004*a*).

nitions can be adopted by distinct types of supply chain. Organic sales in the UK, for instance, are dominated by corporate retailer sales (70 per cent) and overseas (extended) procurement (70 per cent of all retailer supplies). The implication is that regional and ecological definitions are vulnerable to substitution, duplication, and intense competition between extended, proximate, and face-to-face chains themselves. Second, and perhaps more positively for smaller and local networks of producers who are attempting to capture more value through 'shortening' chains and redefining quality around sets of local-ecological and bio-regional criteria, the evidence suggests (see Allen et al., 2003*a*; Murdoch and Miele, 2004; Sage, 2003; Tregear, 2003) the complex evolution of the social and economic diversity and fission in producer–consumer relations within the alternative sector. While the conventional sector is also rapidly developing product differentiation— often on the basis, as Boyd and Watts (1997) remind us, of relatively cheap oversupply of industrialized inputs and related 'surplus-vents'—this is a different pathway from the process of retailer-led standardized differentiation.

In the rest of this chapter we attempt to explore these distinctive geographical dynamics of 'alternative' food networks. We argue that they are based upon a set of spatial and ecological features which play themselves out differently in different regions and define their own competitive spaces within the same region. We will be exploring in more detail the 'clash of paradigms' between the alternative and the conventional sectors in succeeding regional case study chapters, which focus on California, Tuscany, and Wales. Here we will conclude by addressing some of the key geographical dynamics of these alternatives networks with reference to cases which have developed, in North America and Europe, in the midst of the conventional agri-industrial paradigm.

Constructing Value-Capture: Towards New Economies of Quality?

A distinctive feature of alternative networks concerns 'value-capture'. This requires that new social networks and entrepreneurial initiatives are merged with respect to ecological, human, social, and manufactured capital. It also requires that the disposal of the new wealth that is created shows a careful balance between satisfying consumption needs and maintaining reinvestment levels that will assure the long-term future of ecology, the networks, and the enterprises. Overall, then, sustainable wealth creation and local economic development require new entrepreneurial initiatives that focus on investing in the local environment, creating and strengthening local institutions, and employing people and their resources.

In the alternative sphere we can postulate that value-capture at the producer end of food supply chains has at least three potential dimensions. First, producers and their networks attempt to capture more of the *economic value of their products* in a prevailing context in which much of this value is lost to the downstream sectors (Marsden, 2003; Renting, Marsden, and Banks, 2003). Second, as we will outline below, this requires *innovations in the mechanisms for distributing value* among producers and processors. This involves new types of socio-ecological entrepreneurial activity based upon distinctly different types of networks. Third, these two types of value-capture can lead to new potentialities with regard to *forging synergies* between agricultural practices and different types of multifunctional activities such as agri-tourism, engagement in off-farm activities, and environmental schemes and projects. As a result, alternative food chains can also stimulate *multifunctional forms of value-capture*. To increase the possibilities for such value-capture to occur, new local networks and new forms of 'ecological entrepreneurship' become critical, not just in initiating these new valorization processes, but also in protecting and sustaining them against significant countervailing forces.

Such innovative local and regional forms of development need to be seen in the context of two major countervailing forces, within which local 'value-capture' has to fit: globalization and agri-industrial modernization. First, against the backdrop of globalization, where global companies account for an increasing proportion of production and exchange, the very idea of a local economy may seem anachronistic (Ekins, 1997: 19). Yet, despite the threat to economic sustainability, social equity, cultural diversity, and ecological integrity that globalization poses for local communities, many believe that subsumed within this global transition is a strong justification for encouraging the development of local economies. In fact, while global competition—through rationalization of production sites and techniques as well as market operations—offers certain important comparative advantages, this process tends to distribute costs and benefits unevenly across different spatial, temporal, and social domains. Hence, communities that are not fortunate enough to be located on the benefit side of the agri-food global logistics scale tend to experience economic, political, and social marginalization. Local economic development, therefore, can provide an effective counterforce against the forces of global competition. Moreover, as we shall see, it is often those regions that have traditionally been regarded as 'marginal' (i.e. that were never fully 'modernized') that now begin to display the most conducive conditions for the development of alternative agri-food networks and new forms of value-capture.

Second, with respect to rural economies in particular, we have seen that there has been a widespread application of a particular agri-industrial modernization process (which, by and large, is still continuing). This process involves economies of scale and cost-price reduction in the producer sector, further intensification, specialization, and a drastic reconstruction of the rural area so as to create the most favourable conditions for maximizing agricultural (and standardized) production volume. Although this process holds considerable crisis tendencies, it has been further encouraged by logistical retailer-led supply chains and standardized quality regulation (Smith et al., 2004; van der Ploeg, 2003).

These two sets of conditions provide a 'prevailing landscape' in which alternative networks have to be placed. As we shall see, the future long-term success of these food networks depends upon both the robustness of their internal mechanisms and the degree of interaction or boundedness with these prevailing external trends.

Contingent Local Economies and Sustainability

It is important to consider 'the local' in this context as *a form of social contingency*, that is, as a space for rearranging possibilities that attempt to counter the prevailing forces in the agrarian landscape. 'Local' then becomes potentially a social space (*a place to share some form of disconnection*) for the reassembling of resources and value; a place for evolving new commodity frameworks and networks; a place of defence from the devalorization of conventional production systems.

As actors in their own right, local economies offer their own brand of comparative advantages. Through network building, local human capital—knowledge, skills, creativity, commitment to community, and a shared vision of the present and the future—can be harnessed to build and cement mutually beneficial relationships among suppliers, producers, and consumers. A sense of shared ownership of community resources and the responsibility for its viability and preservation 'can inspire trust and commitment, effectively lowering transaction costs and facilitating the process of economic interaction' (Ekins, 1997: 19), without marginalizing social and environmental capital.

Marsden and Smith (2005) outline two case studies in which the problem-solving aspects of partnership-building at the local community level, and a reliance on local capital,[1] have developed to mitigate, if not reverse, several of the negative consequences imposed upon two local communities as a result of the globalization and modernization of agri-food production and markets (see Boxes 1 and 2). For both these local entrepreneurial networks, located in Wales and the Netherlands, sustainable development in the wider sense, rather than merely sustainable economic development, was a major motivating factor.

Along with a number of other scholars, Roch, Scholz, and McGraw (2000) reject the contrasting ideas that, first, individuals have full autonomy over the acquisition and use of information, and, second, that available information, beliefs, and values are fully determined by the prevailing social context. In fact, they argue that 'while the social milieu constrains the range of alternative discussants available to an individual, it also provides opportunities for the individual to meet and consult with new discussants' (p. 778). This relates to the trajectories of problem-solving network building which led to the Graig Farm and Waddengroup Foundation developments (Boxes 1 and 2). In both cases, an effective operating milieu was created in which innovations could thrive. One central part of the construction of such milieu is the development of a new form of 'ecological entrepreneurship', whereby key actors are committed to preserving cultural, ecological, and environmental integrity while also finding new pragmatic ways to create economic benefits (e.g. employment) in the local community. This involves the risky identification of potentially high-value traditional products and practices as well as the use of new regulatory and legal structures (e.g. logos and trademarks) to develop and protect niche products.

These network-based forms of *ecological entrepreneurship* can foster the wider development of 'socio-technical niches' (van der Ploeg, 2003) that can be seen as collective attempts to resist the dominance of the globalization and modernization processes. In this context, we argue, it is important to examine not only the networks themselves but also their substance and social

---

[1] This includes funds, knowledge, skills, labour, commitment, and so on.

Box 1. The Graig Farm network in Wales: organic production and sustainability

The Graig Farm network promotes social, economic, and environmental sustainability from a number of perspectives:

(a) *Farmers' advantages*

- Producers become integral parts of a network that functions through group meetings, invited expert talks, and farm visits, thereby improving the knowledge that allows them to farm the way they always wanted to farm. These frequent opportunities to meet and discuss individual as well as shared problems have facilitated both knowledge-building and problem-solving.

- The levels of trust engendered within the network make it easier for certain productive resources to be shared among its members.

- With Graig Farm acting as the central marketing agent for the group, producers are spared the cost and effort of having to plan and execute individual marketing programmes. Hence, there is an opportunity to concentrate, rather than fragment, farm resources, with each party focusing on what it does best. Farmers with finished lambs, for instance, will notify Graig Farm, which makes every effort to match the supply with market demand—through its farm shop, mail order retailing, a chain of independent retailers, the multiple supermarket chains or via export.

- The producers/marketer partnership allows farmers to have instant feedback on the quality of their animals and on any change that may be required to improve specific quality standards.

- Farmers are assured of a reliable market for their livestock at fair prices.

- Significant developments have taken place with regard to the traceability of products from the farms to the point of consumer purchase. Label and barcode systems are used at each stage and maintained as products pass through the various stages of processing at Graig Farm. The identity of each farm is kept on the labels, and information on each farm can be found. Welsh Black Cattle meat is a main speciality, and specified butchery techniques, including vacuum (biodegradable) packaging, have developed. A team of skilled butchers break down the carcasses into retail-sized packs. Orders can also be freshly butchered to customers' requirements.

(b) *Graig Farm benefits:*

- By working as part of a network, any problem of quality can be communicated instantly to the producer of each animal, thereby reducing the likelihood of small problems becoming systemic problems with significant long-term consequences. The same applies to risks associated with any deviation from the approved organic standards.

- The partnership approach to future production planning allows Graig Farm to be assured of a continuity of supply and quality to meet customer demand; this is positive for both producers and the marketing agent.

(c) *Consumer benefits*:

- In an era dominated by food scares, consumers can have confidence in the organic farming system, which, by law, requires adherence to prescribed production techniques.

- The Graig Farm network facilitates easy traceability of organic meats through personal knowledge of the farms and farmers and farmers' personal knowledge of each animal they rear. This is due to predominantly local sourcing.

- With no external middle-men involved in sales that pass through Graig Farm's farm shop and by mail order, (local) customers can enjoy prices that are as low as possible without negatively affecting producer margins.

Through the development of Graig Farm and the Graig Farm Producers Group—assisted by knowledge borrowed from tropical agriculture and the quality standards and economic support that have been available to UK organic farmers—many livestock farmers along the English/Welsh border of mid-Wales have been able to mitigate the encroaching economic crisis in conventional UK agriculture. This has been assisted by the deliberate diversification of marketing outlets and the corresponding independence from supplying the main corporate retail chains (see Fig. 1). In fact, this partnership has been so successful that, with economic prospects constantly worsening for conventional livestock farmers in the area, the number choosing to convert to organic production and become members of the Graig Farm network has increased dramatically from 2 in 1990, through 20 in 1999, to over 200 in 2004 (Banks, 2001; Smith, 2002).

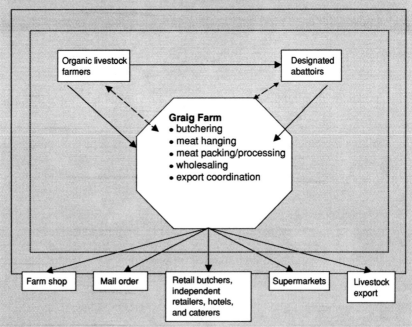

Fig. 1. The Graig Farm network

Box 2. The Waddengroup Foundation in the Netherlands: quality production and location branding

Like agriculture in the UK, Dutch agriculture epitomizes the modernization-productivist trajectory. Production efficiency and cost reduction have, to a large extent, been achieved through specialization, intensification of production, scale-enlargement, and a philosophical reconstruction of the countryside into a large 'agriculture factory' (Roep, 2001). Over time, however, the notion that persistent modernization and rationalization would keep Dutch agriculture globally competitive came under severe stress as global markets continued to show an increasing appetite for ever cheaper products. Those farmers and Dutch regions that were unable to remain viable participants in this agricultural 'race to the bottom' soon found themselves marginalized. This was the scenario that preceded the Waddengroup Foundation initiative.

The seeds of the Waddengroup initiative were planted in 1976, when the van Rijsselberghe family, owners of the Sint Donatus farm on Texel—the largest of the Dutch Wadden Islands—attempted to start the first organic farm in the Netherlands. This pioneering attempt to forge an economically viable and ecologically friendly *disconnection* from conventional agriculture encountered many obstacles and challenges, especially during the early years. However, after encouraging successes in producing and marketing what was branded 'Texel Environmentally and Nature Friendly Products', it was realized that there was an absence of critical mass to make a real impact in the market place. Marc van Rijsselberghe, in 1994, catalysed a network approach to solving this and other related problems that many of his colleague Wadden Island farmers shared (many of whom had already started to produce to organic standards). These shared problems included:

- a sharp reduction in the number of farms and farm employment in the area;

- declining incomes and outward migration;

- significant environmental losses (especially of an uncharacteristic Dutch landscape of leafy hedgerows) due to scale-enlargement farming; hence, loss of spatial diversity and places of specific natural beauty and the loss of traditional breeds and architecture;

- standardization of products for world markets and ever declining prices were leading to the loss of 'traditional' ways of producing, processing, and consuming within the Wadden Islands.

The cornerstones of the Waddengroup initiative were:

- combining local experiences and effort to build up a collective capacity in producing primary products (Texel sheep and a variety of cheeses, for instance), in processing, distribution, and sales;

- using collective knowledge to support new members and others engaged in related businesses within the Wadden area;

- implementing, by means of a registered trademark and a common logo, a collective presentation for a wide assortment of products from the area on

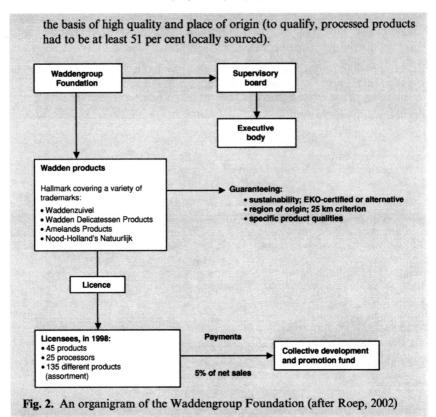

the basis of high quality and place of origin (to qualify, processed products had to be at least 51 per cent locally sourced).

Fig. 2. An organigram of the Waddengroup Foundation (after Roep, 2002)

ecologies. In particular, the fact that they have to contend on a dynamic basis with strong and often countervailing competitive forces puts more emphasis upon the new entrepreneurial abilities of the network members. There is, therefore, a set of interrelationships among network building, the exploitation of production, and marketing synergies based upon quality foods and the new spatial development of socio-technical niches. This process parallels Allaire's conception of the new economy of qualities and the development of what he calls the 'identity paradigm'. He argues (2004: 86):

within the identity paradigm, for example, origin cannot be reduced to the physical characteristics of the product. Transcendant resources not only have strategic power, but as rational myths they offer a cognitive structure to integrate quality. Philosophical principles are embedded in organic food, aesthetics in natural food, solidarity in fair-trade food, and so on. These immaterials are incorporated by products through images that ultimately make these identity resources recognisable in the form of product networks engaged in the economy of signs.

Several writers have questioned the assumptions associated with 'spatial valorization' and the potentially unproblematized and romanticized links among local foods, quality, and sustainability (DuPuis and Goodman, 2004; Goodman, 2003; Winter, 2003). Hinrichs (2003), for example, explores these processes by focusing on the significance of spatial embeddedness in Iowa (see Box 3). In this context, it is important to recognize that 'being local' is not a sufficient prerequisite *in itself* to engender the sorts of innovative frameworks we analyse in this book. As Holloway and Kneafsey (2000, in Winter, 2003: 30) have argued: 'the valorisation of the local may be less about the radical affirmation of an ethic of community or care and [have] more to do with the production of less positive parochialism and nationalism, a conservative celebration of the local as the supposed repository of specific meanings and values'.

Such conceptualizations of 'defensive localism' (Winter, 2003) are clearly relevant in certain social and cultural contexts. However, with regard to agri-food, what matters is not just the 'label' of local but, rather, (1) how the local/ regional is constructed and used in relation to new forms of economic and social networks, which in turn provide a basis for innovation and new types of economic development; and (2) how these new spatially based networks set up and continue to demarcate their spatial and competitive relations and boundaries with the conventional food system. Moreover, what characterizes these types of novel development has only partly to do with the fact that they are producing a particular and more locally grounded type of food. From a conceptual standpoint, they represent new forms of more (ecologically based) social organisation that link producers with consumers both within and across spaces, while also 're-rooting' (as well as re-routing) these supply chains in particular spaces.

The creation of new quality food 'spaces' is then in need of further critical attention, not necessarily in terms of presenting additional case studies, as illustrated in our Boxes 1–3, but in terms of conceptually identifying what is distinctive, socially and economically, in their evolutionary and highly competitive development. In other words: do the emerging quality food spaces begin to represent the evolution of a more sustainable rural economy—a paradigmatic shift—based around the redefinition of social, economic, and agri-ecological resources? Or are they destined to remain niches among a wider economy which continues to devalue local and rural natures and, as a result, *aspatialize* rural space?

Reclaiming Sustainable Rural Spaces through the Development of Alternative Food Networks

As part of our analysis of previous cases (Marsden and Smith, 2005), we believe it is important to assess how alternative food networks actually facilitate 'value-capture' for producers involved and to demonstrate (rather than just assert) that this value is innovatively shared among the producers

Box 3. Relocalization in the US agri-industrial heartlands: state-wide relocalization in Iowa and Minnesota

In Iowa, the increasing realization of deterritorialization by farmers and consumers has led to a plethora of local food initiatives since the mid-1990s (Hinrichs, 2003). Organic agriculture increased from 13,000 acres to 150,000 acres between 1995 and 1999. Farmers' markets and Community Supported Agriculture (CSA) schemes have also mushroomed. A strong emphasis has been placed on sourcing and buying 'locally', which usually means state-wide.

Producers were the first actors to begin these new initiatives. They were responding to the cost-price economic problems of the agri-industrial system, which severely affected independent farmers from the 1980s farm crisis onwards. County extension offices and local chambers of commerce were also active. There were 50–60 farmers' markets in the early 1980s, increasing to 120 by the mid-1990s in a state with 3 million people. CSAs, whereby consumers support the producer by paying for their shares of the yearly production in advance of the season, have emerged more recently. By 1996, there were nine CSAs in Iowa.

Both farmers' markets and CSAs involve a small number of producers and remain highly fragile. This fragility is not just associated with the precarious consumer demand, especially in localities beyond the university towns, but also with internal ethical and moral dilemmas of the participants (Hinrichs, 2003).

Two other initiatives have involved (1) a publicly funded demonstration project on public procurement of local foods; experiments of hospital and university procurement of local foods have taken place with the intention to maintain small and medium-sized farm holdings at the same time as increasing the circulation and profile of local foods; (2) promoting and marketing the cuisine of the Iowa-grown banquet meal, a socially reconstructed Iowa meal that started in 1997, at the tenth anniversary of the Leopold Center for Sustainable Agriculture—a research centre based at the land-grant university and funded through taxes on agri-chemicals. Set explicitly as a state-based antithesis to the market power of the conventional chains, this meal comprises traditional pork and vegetables. By 2000 the Practical Farmers of Iowa had coordinated 47 different events drawing on a network of 23 Iowa farmers. The meal has become more embedded in the restaurant culture of the state and has been helped with some state aid. Hinrichs (2003) argues that the consumer's decision to eat the meal represents a self-conscious departure from the deterritorialized chains and that the very act of 'the meal' begins to break down the constructed distantiation between producers and consumers in the conventional chains.

In Minnesota, as part of the National Campaign for Sustainable Agriculture, the Crossroads Resource Center, based in Minneapolis, is a pioneer in analysing local farm and food economies. It has written an innovative business plan for a cluster of regional investment funds in Minnesota and is lobbying state officials to enact supportive state legislation. These organizations point to the growing imbalances between food production and local value capture and local community development. For instance, situated in the heart of nearly a billion dollars' worth of annual (conventional) food production, the town of Houston recently spent two years without a food store (Meter and Rosales, 2001). In the Hiawatha's Pantry Project, it is estimated that approximately $800 million leaves the region's economy each year as a result of the perverse conventional system. The intense cost-price squeeze is illustrated in the south-east of the state, with the 8,436 farmers gaining an aggregate income of $866 million, but having an overall cost of production of $947 million. They argue that the erosion of the farm economy is affecting the whole region, which is losing its capacity for ownership. Farm household incomes fell by 40 per cent between 1990 and 1997.

Local food initiatives are heralded as a major tool in reversing these race-to-the-bottom tendencies. Root River market in Houston was developed by nine residents in 1999, when they formed a locally owned cooperative grocery business after the disappearance of the town's last store. Some 350 locals joined the cooperative, which then came to include a majority of the adult population (total pop. 1,000). Each member purchased a $100 share. Other investments were added and the cooperative managed to open a new store of 8,000 square feet. The store places emphasis on healthy food choices and local produce.

Forty miles to the north, Rebekah's Restaurant has been opened in Plainview. Paula Wheeler and Diane Lutzke buy food from 13 local producers purchasing more than $15,000 of fresh food annually. All the meals are cooked on site and are seasonal. The attraction and success of the restaurant lies in the use of local organic supplies. Close by are also several CSA initiatives, such as the Earthen Path Organic Farm and the Full Circle Cooperative (Oak Center). Led by one of the pioneers of organic food in the state, Steve Schwen, there is a network of 14 local organic producers which conveys products to local stores and restaurants. Farmer cooperation is seen as critical to the continued expansion and success of the network. An average of $800–1,000 of produce is sold each week during the seasonal cycles (roughly from mid-May to December). The search for new members to extend the production base is a key feature. They are committed to building a 'sustainable region' and to redesigning producer–consumer relationships from a strong organic presence (Meter and Rosales, 2001).

and the local buyers in different types of partnership arrangements. In the Graig Farm case (Box 1), for instance, in 1999, there was a 50 per cent premium on large lambs and a 100 per cent premium on small lambs for members of the network. Moreover, by comparing conventional livestock auction market prices for beef cattle with those received by Graig producers, we identified a premium of 27 per cent in some cases. There was some evidence that this was not just associated with the traditional premium on organic lamb and beef, given that conventional prices through Graig also showed a premium compared with conventional prices associated with the traditional livestock markets. The producer network has grown significantly since 1999 as a result of the relative economic attractiveness (not least in the reductions in transaction costs) of supplying through Graig and the protection from the further falls in farm gate prices in the conventional sector. Indeed, one economic advantage is the creation of more stability in farm prices as long as partnership arrangements can be effectively maintained. There is, therefore, an active and dynamic process of value-capture occurring, which is socially and spatially based.

In the Waddengroup case (Box 2), being part of the network implied a range of new opportunities for participating farmers. The new marketing channels, the protection of the hallmark, involvement in new networks and an increased flow of clients offered considerable potential to consolidate and generate further income, thus broadening and deepening agricultural production. Several new on-farm activities can lead to more solid forms of multifunctionality. One case study farm reached extra returns of 21 per cent (Roep, 2002), mostly over and above the usual premium for organic production. If the extra activities (such as agri-tourism and environmental schemes) are included, the extra returns reach 44 per cent. It is the food supply chain links, however, which contribute most significantly to extra returns. An analysis of farm accounts shows significant premiums deriving from being involved in the network. In the case of the Sint Donatus farm, for instance, returns from food supply chain innovations reached 335 per cent in comparison with conventional producers through value-adding processes, especially in dairy products. This case offers full-time employment to 10 people, compared with 1–1.5 on the average dairy farm of the same size (i.e. the same milk quota). Such a difference underlines the significance of enabling 'economies of scope' through network development and the financial opportunities of creating synergies among food value-capturing activities, direct selling of farm products, and agri-tourism.

Both case studies show that the problem-solving capacity of networks can be enhanced through the presence of entrepreneur-type facilitators and the *openness* of the network to ideas that originate from outside as well as inside the action milieu (Roch, Scholz, and McGraw, 2000). While sustainable development in local communities will depend on how successfully local capital (funds, knowledge, labour, culture, a shared vision, and environment) can be merged with local entrepreneurial ability towards agreed objectives, much can also be learned from the external community to enhance local initiatives. Hence, these cases (both in Europe and the US) are not just about new forms of localism. Rather, they display ways in which relocalization can contingently create spaces which bring together new assemblages of local and external knowledges and practices.

In the cases summarized here (Boxes 1, 2, and 3), the problems of economic marginalization and creeping rural decay (not least linked to the continuing crisis in agricultural modernization and its policy frameworks) have been successfully arrested by the move to quality farming and food production, facilitated, in some cases, by organic farming standards. Hence, while the problem of sustainable rural development is by no means solved in these localities, important progress has been made.

At a more conceptual level, the development of these alternative and locally derived networks raise some important questions concerning the degree to which they are economically and socially sustainable over time and how they could become more diffused over larger areas of rural space as the crisis in agricultural modernization continues. For instance, wider research conducted across six European countries (see van der Ploeg, Long, and Banks, 2002) estimates that up to 50 per cent of farmers are, to varying degrees, following broader or deeper rural development strategies, with many of them *combining* these with continued participation in conventional agricultural markets. While the two European examples represent particularly well-developed counter-movements to conventional and more aspatialized agri-food systems, they are emerging in a more widespread fashion, thereby increasing the total amount of locally dependent initiatives (see Renting, Marsden, and Banks, 2003). These developments are, however, much more 'hidden' from official data sources, relying, as in our Minnesota example here (Box 3), on alternative research. They are, in general terms, a response to the recent crises in conventional agricultural costs and prices and, at the same time, opportunistic and entrepreneurial attempts to capture more value-added from a larger segment of quality-seeking customers.

More conceptually, we can see these new networking activities as distinctive attempts to reclaim parts of the rural land and its social space back from the homogenizing tendencies of the conventional system (Hinrichs, 2003; Marsden, 2003; van der Ploeg, 2003). As van der Ploeg (2003: 379), echoing Kautsky (1988) a century earlier, argues:

a particular ordering of space is implicit in all labour and production processes. Different farming styles result in different spatial constellations, just as a particular spatial constellation lends itself to certain development patterns and hampers others. Not for nothing is the struggle for accelerated scale enlargement translated into the compartmentalisation of rural areas, in the creation of 'free-havens' or 'enclaves'.

In one sense, these alternatives demarcate in one sense such 'free-havens'; that is, areas in which new social and environmental landscapes begin to take shape and are released from the traditionally regulatory 'grip' of the conventional system. In the examples outlined here, both in Europe and in the US, the new agri-food developments have also spawned new labour and community practices, which then contribute to further economic capture of value for the regions. In this sense, it is possible that the social and economic reach of such agri-food developments can be far greater than the some of their parts, creating new capacities in a more diversified rural landscape.

There are, of course, significant tensions in this quality and spatial battleground, given the continued dominance of conventional chains and their attendant and competitive regulatory systems. In the UK, the onset of the foot-and-mouth crisis in 2001 and the dominant role of the corporate retailers in both procuring and selling over 70 per cent of organic products highlight the strong tendency of the conventional 'cost-price squeeze' to be replicated in the organic sector (see Smith and Marsden, 2004). This is a major concern for networks such as Graig Farm, which in part draw their economic strength from the diversification not only of farm production but also of their retail markets. Overdependence on the large retail multiples holds considerable dangers for them; in fact, their large slice of the organic market in the UK means that they play an influential role in overall organic product price setting. It is important to recognize, however, that the conventional retail-led chains hold a different economic and social relationship with the local and regional landscape, compared with the alternatives described. In fact, conventional chains are concerned with extracting value from it, rather than capturing value for it.

This is a significant difference in the spatial relations between conventional and alternative agri-food chains. In the cases we have analysed, there is a competitive co-production of the wider rural landscape, with the valorizing processes associated with conventional production lying contiguous to the more heterogeneous and revalorizing processes associated with quality-based networks. Many farmers in the new networks are also practising in both systems. In one sense, this is the new value-based dualism affecting producing regions, one that influences the social, economic, and environmental aspects of their rural landscapes.

In another important sense, however, and partly as a consequence of the arguments outlined here concerning divergent landscape 'capture', the new networks need more spatialized (rather than sectoralized) institutional involvement. A potential barrier to their development, and a reason why their diffusion is hampered, concerns the domination of *competitive forms of conventional regulation*, or aspects of regulation associated with competition policy, food safety and hygiene, environment and planning, and the private forms of regulation increasingly implemented by corporate retailers. As we shall see in the case study chapters in the book, alternative food networks need alternative forms of regulation and support that give legitimacy to their initiatives. This can occur through R&D and marketing support, for instance, but support is also needed to *defend* rural spaces and niches from the devalorizing tendencies of the old style, corporatist and clientelistic CAP policy instruments.

As Brusa (2003) has convincingly argued with respect to such new forms of local development in southern Italy, rather paradoxically, it has traditionally been the actual *degree of distance* that producers and processors can create from the CAP and its associated regulatory systems that influences the degree of success in creating new quality networks. This is a distance from the *lock-in* effects of production-based subsidy structures, intensively based production systems, low value-added chains, and often traditional corporatist and clientelistic farmer–farm union–state relationships, which in themselves still derive power from lock-in. In both the EU and the US, to differing degrees, the role of the state remains critical in providing opportunities and in creating barriers for alternative networks.

In order to develop, therefore, new networks not only need to create alternative internal quality assurance systems; they also need external institutional support to *assure and defend their spatial and social boundaries* which *can sustain the benefits of 'lock-out' from the 'prevailing landscape'* and bolster different types and packages of technologies and techniques. We see this regional institutional development, for instance, in Italy around speciality cheeses and wines and organic production (de Roest, 2000).

In Wales, the Netherlands, and the US, given the stronger application of the conventional system along modernization lines, such regional and local strategies are starting from a much lower base. As we shall see in Ch. 6, in Wales the Agri-Food Strategy and Agri-Food Partnership were established in 1999 to provide a territorial approach to agri-food through the Welsh Development Agency. Such an approach sits alongside the conventional (sectoral) allocative system of the CAP.

Our case studies also highlight that the success of alternative food networks rests with new forms of associational involvement not only among producers but also along the newly formed supply chains themselves. These very often come together through what we might term *backward–forward technological innovations* or *retro-innovation*, such as old butchery and slaughtering practices, curing techniques, and 'traditional' forms of pest management (Stuiver and Marsden, forthcoming). Alternative food chains rely upon new types of spatiality and agro-ecological relationships (Sevilla Guzmán and Woodgate, 1999) and they combine the use of new and old technologies, as demonstrated by communications to their customers through their websites about the 'traditional' nature of some of their production and processing techniques.

Conclusions: Capturing Spaces, Creating Opportunities

Despite considerable obstacles and constraints, not least from the persistence of competitive regulatory and agricultural policies which continue to 'lock-in' producers into providing standardized food products at ever cheaper farm gate prices, new and highly uneven network developments in agri-food are diffusing and contributing to a more diverse rural landscape in both Europe and the US. This raises important conceptual questions concerning the capacity of local places to sustain these 'counter-movements' and to continue to promote their cause. We have identified some of the key internal and external components that are shaping these new spatial relationships. Embodied in these is also the recognition of a new form of 'ecological entrepreneurship', whereby key actors play a decisive role in enrolling and mobilizing other actors into the network, so as to create and sustain its structures and develop new interfaces between producers and consumers.

We can postulate that this may be an important element in the progression of an agrarian-based ecological modernization (Marsden, 2004a) which raises the need to view ecological entrepreneurship as more than simply an oxymoron in the environmental policy literature and debates. This trend also takes us beyond the realms of generalized 'social capital' justifications for local rural and regional development, suggesting the need to match an understanding of new forms of network development with ecological entrepreneurship on the one hand and with the wider social and political economy of rural and regional landscapes on the other.

While the scholarly literature on alternative food movements and networks has expanded rapidly over recent years, our critical analysis here suggests that more conceptual effort is needed to understand the distinctive geographical and social components of these trends. In particular, *spatial contingency and capture, the degree of disconnection from conventional systems* (i.e. 'lock-in' and 'lock-out'), *retro-innovation,* and *ecological entrepreneurship* would seem to be salient areas for exploration if we are to assess the real sustainability of the new and distinctive agri-food geographies that confront us.

Around the themes of deterritorialization and reterritorialization, this chapter has attempted to provide a conceptual basis and a starting point for studying the regional worlds of food. Interestingly, both the agri-industrial model—shaped by the vibrant integrated cluster developments on the one hand and by the internationalizing tendencies of corporate retailing on the other—and the more fledgling and heterogeneous alternative model are played out increasingly at the regional level. However, as we have tried to delineate in this chapter, they are subject to quite distinct, and some might say contradictory, spatial principles. These are principles of co-evolutionary capitalist and state-based spatial development that collide and compete in and through different rural regions. This is not simply a top-down process; to explore its spatial forces and frameworks in more depth, it is necessary to enter the fluid regional worlds of food we shall analyse in the following chapters.

4

Localized Quality in Tuscany

Introduction

With its rolling hills, small farms, diverse products, and high-quality foodstuffs, Tuscany easily conjures up a world of diversification and localization. In fact, so many of the region's products are seen as world class—notably its wines, olive oils, cheeses, and processed meats—that it is tempting to see this region as the prime example of an Interpersonal World (in Salais and Storper's terms). Yet, Tuscany's perceived success in this world of food is a recent phenomenon. Until the 1990s the region was thought to be rather 'backward' in character, mainly due to its inability to adopt conventional industrial approaches to food production and processing. While some effort was made to shift Tuscany on to a more industrialized development path during the 1960s and 1970s, by the early 1990s this was widely regarded as having failed. Out of this failure, however, came the search for a new development model, one that could work with, rather than against, the region's core assets—notably, its localized variety in foodstuffs and environmental features. Thus, a distinctively Tuscan approach to the agri-food sector is explicitly identified in the recent Rural Development Plan (RDP) drawn up by the Tuscan regional government. The document states that the strategy elaborated in the plan is aiming at 'strengthening the "Tuscan model" of agricultural and rural development'. The plan goes on to identify key characteristics of the model, including the presence of small and medium-sized farms, the existence of quality products, the diversification of agricultural production, the provision of adequate marketing networks, and the enhancement of the environment and the agricultural landscape (Regione Toscana, 2000).

It is tempting to imagine that the consolidation of a diversified and localized world of food production in Tuscany owes much to the implementation of this model by governmental authorities in concert with other actors in the food sector. However, it will be argued below that the emergence of a new world of food in Tuscany owes as much to happenstance as it does to the conscious agency of differing institutions and organizations. For instance, Tuscany's aesthetically valued landscape has remained intact mainly because a system of sharecropping that emerged in the medieval period was maintained until the end of the Second World War. This system was deeply entrenched in the Tuscan countryside: it sustained the dense network of farms and farm buildings, a wide range of crops and animal products and a traditional rural social structure. While sharecropping withered away in the post-war period, it ensured that the Tuscan landscape and its rich variety of food products remained available to later generations of farmers and consumers. Importantly, there was nothing about this process that was planned in advance; rather, the system of agri-food production that had been consolidated in Tuscany fitted neatly into the new world of consumption that emerged in the advanced capitalist countries in the closing years of the twentieth century. The assertion of a 'Tuscan model' simply indicates that the degree of 'fit' between Tuscany and the broader consumption context has now been recognized by a variety of food-sector actors who are building a discourse around this model to consolidate Tuscany's place in the new alternative world of food.

In this chapter we document the emergence of the Tuscan world in order to show how its component parts have been melded together within the 'Tuscan model'. We first consider the system of sharecropping that was so central to the region's agricultural structure and show how this system gave rise to considerable diversity in foodstuffs. We then go on to consider how the agri-food sector was restructured in the latter part of the twentieth century. As we shall see, the move out of sharecropping meant that for a time the viability of many farms was threatened by the application of rational and industrial processes of modern agri-food development. In this context Tuscany was regarded as a marginal region, of no real significance to the industrial world that was then beginning to predominate in Europe (notably through the implementation of the CAP). In more recent years, however, Tuscany's diverse and authentic food products have come to be re-evaluated, mainly by a new generation of consumers (often present in the region in the guise of tourists). This process of re-evaluation has allowed the development of the 'Tuscan model' as an alternative to the previously dominating industrial approach.

In order to show how the emergence of a new context of consumption has affected key Tuscan products we provide a number of short case studies below. These cover the two premier products of wine and olive oil as well as a product representative of localized distinctiveness—*lardo di Colonnata* (a form of processed pork fat that is now seen as a regional delicacy). In the case studies we consider how the contemporary re-evaluation of localized variety has affected processes of production and marketing associated with these products. The cases thus give us some insight into the consolidation of a new world of food in the Tuscan region.

After showing how the constituent parts of Tuscany's world were maintained into the current period, we examine how local political agencies have articulated the so-called Tuscan model. A key development here is the emergence of regional institutions with some responsibility for the agri-food sector in the 1970s. Over time these institutions have gradually woven together an integrated approach to agri-food development, one that is formulated with a wide range of regional stakeholders and is therefore sensitive to the needs of local actors. This 'territorial' (as opposed to 'sectoral') approach is tailored to Tuscany's very peculiar social, economic, and environmental needs. It thus plays a key role in integrating further the various elements that make up the world of localized quality to be found in the region.

Building a World of Quality: A Short History of Agri-food in Tuscany

As already mentioned, the Tuscan countryside was profoundly shaped by sharecropping, or *mezzadria*, a system of production that emerged in the thirteenth and fourteenth centuries and lasted until the end of the Second World War. At its height, during the 1930s, the sharecropping system covered nearly half of Tuscan agricultural land and affected almost all aspects of agricultural production. According to Pratt (1994), the *mezzadria* comprised an urban landlord who provided the farm, the farmhouse, and half the working capital, and a rural tenant who, along with his family, provided the other half of the working capital and the labour. While the sharecropper had some autonomy in terms of the labour process, the landlord regulated the flow of investment and oversaw the processing and marketing of products. All the production from the farm was divided equally between landlord and tenant.

The landlords usually held large estates made up of many sharecropped holdings, and the various farms were coordinated from one administrative centre (the *fattoria*). While this system ensured that sharecroppers worked for the landlord, it also ensured that the tenanted farms were regularly maintained (especially as the *fattoria* frequently employed wage labourers for building, storing, and transportation tasks). The system also permitted regular investment in the farms by the landlord, a factor that was particularly important for the cultivation of wine and olives, where there is some considerable time-lag between initial cultivation and subsequent output.

This system of production tied the countryside very closely into the towns. The landlords were usually based in urban areas and they used the sharecropping system to ensure a steady flow of food into the towns and cities. There was therefore no sharp distinction between town and country in Tuscany: an intricate web of relations bound the two spatial zones together. Urban landlords invested in rural estates, while rural tenants delivered part of their produce to urban consumers.

Yet, while the share of production taken by the landlord ensured some connection between the sharecropped farm and the demands of the urban market, for the most part the *mezzadria* was focused upon subsistence agriculture. Production on the farms was very varied in order to allow the whole family to survive through the year. Pratt (1994: 39) describes this variety in the following terms:

most farms kept cattle, sheep, pigs, rabbits, chickens and pigeons. A patch of wood-land provided household fuel for cooking and grazing in high summer. Arable land produced wheat, barley, oats, maize, beans, hay and a variety of other crops to give oxen their working strength. Vineyards and olives were created by terracing the steeper slopes, while figs, walnuts, almonds and a dozen kinds of fruit tree were common. A kitchen garden supplied onions, garlic, carrots and all the other veg-etables and herbs needed for a diet centred on bread and vegetables.

It is easy to imagine this mixed farming system yielding the diverse landscape that still exists in Tuscany. It also yielded countless locally significant food products, especially once producers began processing meats on farm. In short, Tuscan cuisine has its roots in the subsistence production of share-cropped farms.

The longevity of the *mezzadria* system ensured that it profoundly shaped the character of the Tuscan countryside. However, by the time of the Second World War, discontent with the system was growing (despite attempts by Mussolini's Fascist government to re-entrench sharecropping relations be-tween tenants and landowners). This discontent was bolstered by the active role that tenants played in the resistance movement during the war (Brunori, 2005). Thus, after the liberation of Italy in 1945 a political struggle for land reform broke out, orchestrated in part by the Communist Party. As Pratt (1994: 50–1) says: 'the *mezzadri* first wanted to wrestle from the landlords greater control over estate management then a greater share of the product, and finally ownership itself'.

By the late 1940s the Christian Democrat government, faced with wide-spread rural agitation, announced a partial reform programme, which be-came fully operative in 1953. The main aims of the reform were to end the *mezzadria* while at the same time promoting a modernization of the entire farming system. Implicit within the reforms was a desire to shift agriculture from subsistence production to production for the market. This seemed to require a rationalization of the farming structure, an intensification of pro-duction processes and greater specialization in terms of crops grown and animals reared. Thus, the reform agency built new rural roads and set up cooperatives for the leasing of farm machinery, notably tractors. According to Pratt (p. 62), the reform quickly succeeded in increasing the productivity of land in much of Tuscany. Indeed, the number of cattle and pigs more than doubled and the amount of land given over to wine and olives increased rapidly. Moreover, as farmers became better integrated into urban markets, more money became available for further investment, so that by 1961 almost one-third of farms owned their own tractors (Pratt, 1994).[1]

Brunori (2005) identifies two main impacts of the reform in Tuscany: 'On the one hand, reform and rural emigration allowed the growth of a family farm sector, based on land ownership and family labour. On the other hand, landowners were stimulated to capitalise their farms, that is, to make new investments in farm infrastructures and to replace sharecroppers with employed workers.' Both these trends permitted a much-needed rationalization of the sector so that production could be better geared to market demands. However, the size of Tuscan farms remained small—by 1969 the average farm size was still only 29 hectares. As Pratt (1994: 20) points out, 'the farms were always too small for modern forms of mechanisation ... they [were] also too small for the labour available, and this led to a gradual population decline and the search for off-farm employment'. The search for off-farm employment was also encouraged by the development of new economic activities, especially in the areas around Florence, Siena, Pisa, and Livorno, where the growth of new industrial districts provided alternative forms of employment for those struggling to stay on the land. Thus, in many traditional farming areas, especially in hill and mountain regions, farms were abandoned as families moved in search of work in the new burgeoning industries (this was especially the case in areas that had lain outside the *mezzadria* system). As Pratt (ibid.) puts it: 'the poverty and isolation of rural life compared to an increasingly known urban alternative led to a rural exodus'.

This trend was exacerbated by Italy's entrance to the EEC and its adoption of the CAP in the 1960s. As it has been extensively documented, the CAP sought to increase farm productivity across the European agricultural space in order to boost overall levels of productivity. In so doing, it aimed to further rationalize production so that farms could be placed on a sound economic footing. This meant increasing the size of holdings, reducing the number of workers, and encouraging the use of new production techniques. While these objectives were largely in accordance with the goals of the Italian government's post-war reform, they once again implied a marginalization of Tuscan agriculture. By European standards, Tuscan farms were small and inefficient. Thus, it was believed that only by encouraging a rural exodus and a further integration of existing holdings could agriculture be rendered more economically efficient.

For much of the post-war period, then, Tuscan agriculture was seen as lagging behind the more modernized agricultural regions to be found in large areas of northern Europe. While some special provision was made for Tuscany's landscape and terrain, there were continual efforts to render the

---

[1] As Pratt (1994) points out, this figure gradually increased and by the early 1970s few farmers were using the leasing cooperatives.

sector more economically effective. These efforts seemed to imply fewer farms, fewer workers, and fewer products. By the 1980s, the application of this modernization approach, and its failure to shift Tuscan agriculture in the desired direction, engendered a crisis in the sector. As Brunori (2005) describes it, this crisis revolved around cereal and animal production:

the wheat sector was not big nor organised enough to be a valid supplier of the big pasta or industrial bakery companies located in the north of Italy. In the animal production sector, the environmental conditions of Tuscany did not allow a process of heavy technological restructuring towards large-scale intensive farms. . . . The closure of a large number of small abattoirs imposed by the European safety and hygiene regulations in the meat sector and the on-going concentration of retail distribution marginalised meat circuits in Tuscany.

Picchi (2002: 206) suggests that the CAP also failed to support products such as wine and olive oil that were oriented to high-quality market segments. He claims that such products hardly benefited at all from the new policies.

The advent of this crisis appeared to mark the limits of agricultural modernization in the region. Yet, fortuitously, at precisely the moment the limits of the modernization strategy were becoming clear, a re-evaluation of the traditional sector began to take place. In large part, this was due to the emergence of tourism as a key means of generating economic activity in countryside areas.

Foreign visitors have long been attracted to Tuscany. However, by the 1970s, the development of international airports in Pisa and Florence as well as motorway links to northern Europe allowed ever greater numbers of tourists to visit the region. In 1986 Tuscany received more than 6 million visitors, a figure which represented a 160 per cent increase since 1960 (Sonnino, 2003). Traditionally visitors had been attracted to the urban centres to follow art and architecture tours. However, the attractions of the Tuscan countryside gradually drew ever greater numbers of tourists into the rural areas. This gave rise to what has become known as 'agri-tourism', an activity that enables visitors to stay on farms as 'guests' of the farm family. As Sonnino (2003) points out, agri-tourism grew rapidly in the 1980s. Between 1984 and 1988 the numbers of farms providing agri-tourist services went from 108 to 454, the largest increase in the whole of Italy. Moreover, growth in the numbers of agri-tourists gave rise to a 'feedback' effect, as many of the abandoned farms left behind by post-war rural out-migration were purchased by wealthy foreigners, notably from Britain, Switzerland, and Germany. As a consequence, the number of new agri-tourist facilities further increased, and by the late 1990s there were around 1,500 agri-tourist farms in the region (Brunori and Rossi, 2000).

As Sonnino (2003: 132) points out, agri-tourism appeared to provide an ideal diversification strategy for small farms. Moreover, this strategy could successfully revalue intrinsic aspects of Tuscan agriculture, notably the variety of local products and the diverse beauty of the countryside. Tourists are attracted to the Tuscan countryside precisely because it retains so many distinct products and holds such a varied landscape. Once these features are (re)discovered by large numbers of tourists, new markets for forgotten products and services come into existence. The huge growth in tourism in Tuscany (currently, on average, around 20 million people visit the region each year, 11 million of whom are foreigners) provides a new rationale for maintaining the traditional countryside in existence and signals the end of 'modernization', at least as it was practised in the 1960s and 1970s. In short, it lays the basis for the assertion of an alternative Tuscan model of rural development.

Case Studies: Re-evaluating Traditional Tuscan Products

The successful development of the tourist sector in Tuscany owes much to the multitude of local food products that comprise 'Tuscan cuisine'. As we have seen above, many of these products are linked to the *mezzadria* system, which induced farm families to cultivate a variety of differing foodstuffs so as to ensure their own subsistence. These foodstuffs were also supported by locally situated consumers in both urban and rural areas so that local cultures of consumption came into being. However, the small scale of these consumption contexts meant that farmers always struggled to maintain their livelihoods. Thus, efforts were made in the post-war period to modernize the agricultural sector in the hope that farmers could tap into wider markets, both in Tuscany and further afield.

Following the failure of the modernization effort, it became evident that local consumption cultures might be strengthened through tourism—rather than exporting local products to distant markets, consumers could be encouraged to visit the areas of production and to develop a rich and rewarding relationship with both the producers and their products. Given that most of these visitors were from areas outside Tuscany, it was inevitable that the demand for Tuscan foods of many differing types would grow. In fact, certain products have now come to be seen as among the most valued in the world. In this section we consider three such products—olive oil, processed meat (*lardo di Colonnata*), and wine (Brunello di Montalcino)—in order to show how the transformations in the Tuscan food sector that we identified in the previous section have affected traditional Tuscan foods. In briefly describing these three cases, we seek to show how the changed context of consumer demand has facilitated a new approach to agri-food development, one that works *with* the traditions and resources inherited from the past but *modifies* them in line with current requirements. It is this approach that largely defines the Tuscan model to be considered in the following section.

Olive Oil

Olive oil is a traditional product that has come to be associated closely with high quality in new consumer markets. Yet, in many ways it retains a traditional production structure, one that is not so open to the kind of rationalization processes that we will discuss in the case of Brunello di Montalcino. Nevertheless, the re-evaluation of olive oil has enhanced the status of the whole region and, again, it helps to ensure that the Tuscan world of food is seen as a quality world.

As with many other products, the production structure for olive oil can be traced back to the sharecropping system. A high number of the formerly subsistence farms in Tuscany produce olives, often only for household consumption. Olive oil production is therefore widely dispersed throughout the regional territory, with many farms growing the crop in tandem with other crops. According to Belletti and Marescotti (1997), only 23 per cent of the total number of farms growing olives specialize in olive production. Thus, while some olive oil is grown intensively for export, the vast majority is grown on a small scale for domestic purposes. It is this latter component of the sector that will be our focus here.

In part, the production structure for olive oil is determined by the terrain, notably the hills and soil types. For these reasons, much olive oil production retains many traditional features, as Belletti and Marescotti (p. 1) explain:

the peculiar characteristics of Tuscan olive cultivation determine very high production costs, due to the difficulty of mechanising various operations, in particular harvest. Harvest is still widely carried out following the traditional method of stripping off, or beating down and picking ripe olive fruits. This method requires a lot of labour . . . . Such labour can be found only partially within a farm or the family of the farmer (family members and other relatives employed in other sectors during the picking period) and labour needs are satisfied either by an 'exchange of labour' with other farms in the area or mainly by day labourers.

Even though this form of production is labour-intensive (and is somewhat impervious to increased mechanization), it does ensure that the olive oil is almost entirely extra virgin in quality. In other words, it tends to reach the highest production standards.

As Belletti and Marescotti (p. 3) make clear, traditional methods and processes also govern pressing and processing:

olives are sent to the mill in almost all cases by the farmer himself. The milling is done by a very high number of olive-oil mills (owned by farmers, cooperatives, and specialised firms) that in many cases provide this service as their exclusive or main activity. The farmers usually want to maintain ownership of the oil obtained from their own olives, not only because of the differences in yields among growers, but for commercial and 'sentimental' reasons as well. So the processing must occur by lots—keeping separate the lot of olives of each grower—with a consequent rise in the cost of production.

In part, the separate processing of each farm's 'lot' stems from the fact that a large proportion of the oil will be used by the farm family. Belletti and Marescotti calculate that up to 35 per cent of the total amount of oil produced is destined for the grower's table. This indicates that the old subsistence approach is still strongly entrenched in the olive oil sector. As Belletti and Marescotti (p. 6) say, for many farms 'the quantity sold is residual, and it depends on the high cyclic productive performance of olive trees'. In addition, because many of the exchange relations among farmers, millers, and day labourers entail payment in kind, that is, olive oil, the quantity sold is also distributed among a high number of sellers. The structure of production is therefore not easily oriented to mass markets; rather, it has developed on the basis of direct sales. Belletti and Marescotti calculate that at the beginning of the 1990s about two-thirds of Tuscan extra virgin olive oil was sold in this way.

The market for olive oil has therefore been localistic in nature, with many consumers purchasing the oil directly from the grower or mill owner. Moreover, Belletti and Marescotti (p. 7) discern among traditional consumers a 'particular type of consumer, directly related by "extended" family relationship or friendship ties to the holder and often residing in the same production area'. This observation reinforces the idea that the world of Tuscan olive oil is a strongly 'Interpersonal World'.

However, as in other areas of Tuscan agri-food production, new consumption patterns have emerged during the 1990s. Following the growth of tourism, especially agri-tourism, olive oil producers began to encounter non-local (often foreign) visitors who were interested in purchasing their high-grade products. This shift in the market entailed some modification in producer practices. Interpersonal relations no longer constituted the most effective means of drawing consumers to the product. Instead, olive oil sellers were forced to market their products more actively, by using road signs, branding, posters in towns and villages, participation in food festivals, and so forth. Belletti and Marescotti (p. 16) thus conclude that 'habit and simple "geographic proximity" are increasingly less important in determining the purchase of Tuscan olive oil . . . on the contrary, "cultural proximity" is becoming more important . . . . Tuscan olive oil is increasingly becoming less a "traditional" product and more a "typical" product'.

The shift from 'traditionalism' to 'typicality' marks a shift in the market for olive oil, with new consumers (mainly tourists) buying the oil because they reflexively perceive it to be an authentic Tuscan product. However, this shift has taken place without any major transformation in the production and processing structure. The farms and processing facilities remain small in scale and produce only small quantities, a significant proportion of which still goes for domestic consumption. This structure ensures the maintenance of localized quality. It has simply been bolstered by the arrival of new consumers in the areas of production. Thus, a localistic world of food has been safeguarded not by the exporting of products to other areas of the world but by the importing of people from other areas of the world into the local production space.

Lardo di Colonnata

Another sector that has seen a transformation of demand but has also displayed some rigidity in the structure of production is processed meats. Again, it is important to recognize that under the *mezzadria* many households engaged in their own processing of meat; hence, a wide range of meat products has traditionally been available throughout the Tuscan countryside. As in the case of olive oil, the production of these meats has been underpinned by a local consumer culture characterized by strong interpersonal relations between producer and consumer. And, again, a shift in the consumption context has meant that some of these meats have become regarded as premier meat products, much sought after by aficionados of the Tuscan cuisine. However, once a meat attains this status then a tension ensues between its localized quality and the scale of global demand. A leading example of this tension is *lardo di Colonnata*, a type of processed pork fat that has recently come to be seen as a highly prized Tuscan delicacy.

Colonnata is a small village located at the end of a long winding road running through one of the three marble valleys in Carrara. Traditionally, men in the village have worked as wage labourers in the marble quarries located in the valley. However, as Leitch (2003: 443) points out, many of the labourer's households owned land where they were able to keep pigs. One of the by-products from the pigs was *lardo*, or cured pork fat.

Leitch (2003) draws close parallels between the *lardo* eaten by the wage labourers and the marble that they cut out of the hillsides during the course of the working day. First, *lardo*, like marble, is transformed (cured) from its natural state by the craft skills of the workers. Second, marble, due to its qualities of porosity and coolness, is always cited as a key production ingredient. Used during the curing process, the marble allows the pork fat to 'breathe' while containing the curing brine. Leitch summarizes the process:

the curing process begins with the raw fat, cut from the back of select pigs. It is then layered in rectangular, marble troughs resembling small sarcophagi, called *conche*. The *conche* are placed in the cellar, always the coolest part of the house. The majority of these cellars are quite dank and mouldy. Some still contain underground cisterns, which in the past supplied water to households without plumbing. Once placed in the troughs, the pork fat is covered with layers of rock salt and a variety of herbs, including pink-jacketed garlic, pepper, rosemary and juniper berries. Finally, a small slab of bacon is placed on top to start the pickling process, and six to nine months later it is ready to eat. Translucent, white, veined . . . pink, cool and soft to the touch, the end product mimics the exact aesthetic qualities prized in high quality marble.

Leitch describes *lardo* as a 'proletarian hunger killer' (after Mintz, 1979): 'eaten with a tomato and a piece of onion on dry bread it was a taken-for-granted element in the worker's lunch. *Lardo* was thought to quell thirst as well as hunger and was appreciated for its coolness on hot summer days.' However, as with so many Tuscan products, *lardo's* proletarian roots began to be frayed by the interest of new consumers in this exotic form of pig meat. Again, early outside interest was generated by agri-tourism (also by high-class chefs who were using *lardo* in their restaurants). However, in the mid-1990s *lardo di Colonnata* gained wider attention mainly due to the activities of the Slow Food movement, an organization that, as described in Ch. 1, is dedicated to saving 'endangered' food products.

Slow Food's interest in *lardo* was stimulated by a police raid on a restaurant in Colonnata in 1996. The purpose of this raid was to place the restaurant's pork fat under quarantine in line with new European hygiene legislation. This action—which was widely covered in the national press—led to a national debate about such legislation and the threat it posed to local foodstuffs. For Slow Food, *lardo di Colonnata* illustrated the nature of threats facing Italian cuisines. As Leitch (p. 446) summarizes it:

lardo presented an unambiguous test case for new European hygiene rules, which insisted on the utilisation of non-porous materials in food production. Although there are certainly good techniques for sterilising the *conche*, marble is porous and its porosity is clearly essential to the curing process as well as to *lardo's* claims to authenticity. Local *lardo* makers involved in this dispute thus had a vested interest in lobbying for exceptions to the generic rules designed for large food manufacturers. Their interests coincided perfectly with Slow Food's own political agenda, in particular its campaign to widen the debate over food rules to include cultural issues.

Slow Food thus began to run a campaign highlighting the dangers of hygiene legislation using *lardo di Colonnata* as a prime example of endangered local foods. This campaign helped to ensure the continued existence of this cured pork fat. It also broadened out the demand for the product among concerned consumers. However, the increased demand for a product that is naturally restricted by the method of production had an entirely predictable consequence—widespread copying. Large butcheries from all over Italy in effect began manufacturing a product they called *lardo di Colonnata* even though their product had no connection to the village whatsoever. In response, the authentic *lardo* producers began to campaign for their product to be given PDO status. However, they failed in this attempt, although a number of producers had managed to acquire legal copyright of the name. The group then formed an association (Associazione Tutela Lardo di Colonnata) in order to regulate use of the name.

In conclusion, the original *lardo* makers felt that the publicity that had been generated by Slow Food and others concerned about endangered local products had actually further endangered their original version of *lardo di*

Colonnata, in the main because it had provided a stimulus to other economic agents to enter 'their' market. Without sufficient legal protection, the original *lardo* would be lost to a non-local, pan-territorial market. Colonnata would become nothing but a name attached to the product, rather than a guarantee that the product had emerged from its distinctive environs.

Brunello di Montalcino

The abolition of the *mezzadria* and the land reform occurred in the 1950s provoked a restructuring of Tuscan farms so that large numbers of share-croppers became, in effect, owner-occupiers. However, as stated above, the reform also generated something of a rural exodus that led to the abandon-ment of large numbers of holdings. One response on the part of the landlords to this process of abandonment was a consolidation of the old estates so that they could be run as modern businesses using wage, rather than share-cropped, labour. In identifying a new economic role for the estates, the landlords began to look closely at the production of quality wine for a growing consumer market. Wine is a traditional Tuscan product (its produc-tion in the region extends all the way back to the Etruscans) but it has gained a new profile in the changed consumption context of contemporary capital-ism. Pre-eminent in changing this profile has been a new generation of 'Supertuscan' wines which have been developed over the last thirty to forty years. And pre-eminent among this group is Brunello di Montalcino, a red wine of renowned quality.

The territory of Montalcino, which comprises a small, saucer-shaped valley of around 24,600 hectares, lies to the south of Siena. Although agri-culture is a major economic activity, only around half the area is cultivated, with olive groves accounting for around 2,500 hectares, vineyards for 2,000 hectares, and pastures, grains, and other crops for the rest (Hayes, Lence, and Stoppa, 2003). Montalcino has long been known as a wine-making area, traditionally famous for a sparkling white wine called Moscadello. Accord-ing to Pratt (1994), the cultivation of red wine in the area can be traced from the middle years of the nineteenth century, when a group of landlords came together to develop new scientific methods of wine production. Using San-giovese grapes, they began production of a distinctive new wine—Brunello—that quickly came to be seen as a product of high quality. Recognition of this quality meant that the Brunello producers became popular during the later years of the nineteenth century with wealthy Italian consumers.

Until the reform in the post-war period, Brunello was bottled and marketed on a small scale (during the first half of the twentieth century, only around 60 hectares of land were used for the cultivation of Brunello vines). However, during the 1960s the Italian government introduced legislation which both recognized the distinctiveness of the wine and protected its special status. By allowing Brunello to be classified as a DOC

(*Denominazione d'Origine Controllata*) wine, the legislation effectively stipulated that it could be produced only within the commune of Montalcino, using grapes from the Brunello strain of the Sangiovese vine. The acquisition of this denomination stimulated considerable interest in both Brunello and in the region of Montalcino (which also produced other well-known wines such as Rosso di Montalcino). As a result, remaining derelict land was bought up by external investors keen to explore the potential of the wine.[2]

By far the largest of these new investors has been Villa Banfi, which is owned by a company based in the US. Villa Banfi owns around 3,000 hectares of land and 800 hectares of Brunello vineyard. It has undertaken a huge capital investment into its landholding, for instance, spending over $100 million on state of the art production facilities (Pratt, 1994). In making this investment Villa Banfi has also substantially rationalized the system of production. Pratt (p. 139) describes this process of rationalization in the following way:

Villa Banfi has organised production so that no manual labour is done in the vineyards between pruning and harvest. Terrain was levelled before planting and the vines planted in blocks of up to a square mile, which made it economically feasible to use helicopters for all spraying. The helicopters can spray twenty hectares compared to one hectare for a tractor. The vines were also planted in a way which allowed all green pruning in the summer to be done by mechanical cutters set at a standard height, ignoring variations in plant growth completely. The managers claim that this planting system has higher initial costs, but reduces the wage bill for annual cultivation.

Pratt (p. 162) interviewed an estate manager at Villa Banfi who emphasized the importance of precision:

precision was a key goal ... precision in the spacing of the vines, in the height of the fruiting spurs, in the volume of the fertilisers applied and in the timing of the harvest. It was achieved through the best available laboratory which provided a flow of data on the chemical needs of the vines, on the best shoots to select for grafting and on the ripening process of the grapes. Variability was a source of irritation, whether it was found in the unpredictability of manual labour or found in nature.

Yet, nature in the form of the soil and the weather still exercise some considerable influence over the quality of the wine produced (despite the mechanization of the process, Brunello wines do vary from year to year). Moreover, the owners of Villa Banfi saw themselves explicitly innovating within a line of tradition. Although they had perhaps taken the standardization of quality wine production to its logical extreme, there was still a keen sense that the wine itself remained unchanged.

In fact, the distinctiveness of Brunello di Montalcino has now been recognized once more under Italy's Law 164, passed in 1992. This Law reinforces

---

[2] Pratt (1994) estimates that three-quarters of the Brunello vineyards are now owned by firms who have acquired the land in the last twenty to thirty years.

the earlier legislation on DOC appellations by introducing a new category that establishes requirements for the production and sale of wines classified as 'Designation of Controlled and Guaranteed Origin' (DOCG). These wines—which include Brunello—are subject to the strictest controls in order to ensure high quality. As Hayes, Lence, and Stoppa (2003: 15) note, DOCG wines can only be produced from grapes grown in registered vineyards. In addition, the quality of the DOCG wine that is sold on the market must be tested and certified before bottling. Each bottle of a DOCG wine must bear a government stamp with an individual number.

This production standard is administered by a Brunello Consortium (Consorzio del Vino Brunello di Montalcino), an association that includes almost all producers of the wine, both small- and large-scale producers (despite levels of external investment in Montalcino, a number of traditional small-scale producers remain). According to Hayes, Lence, and Stoppa (2003), the Consortium 'owns' the Brunello brand: 'it is legally empowered to maintain the registry of vineyards entitled to produce such wine; [it] enforce[s] production and quality standards; prevent[s] illegal imitation; and provide[s] in general for the care, improvement and promotion of the brand'. Thus the Brunello Consortium provides yet another mechanism for raising the profile of the brand and ensuring that it comes to be perceived as a high-quality product (according to Hayes, Lence, and Stoppa about 60 per cent of the Consortium's $1.1 million annual budget is spent on promotional activities). Despite the innovative changes going on at the level of production in the Brunello vineyards, quality criteria are rigidly enforced.

The emergence of 'Supertuscan' wines such as Brunello helps to promote Tuscany as a 'quality' region on food markets. This promotion works not only for wines, but also for other products in the local area. For instance, as Brunori and Rossi (2000: 410) note, the emergence of these well-known brands in the wine sector leads on to the construction of 'wine routes', which they define as a 'sign-posted itinerar[ies] through a well-defined area (region, province, denomination area), whose aim is the "discovery" of the wine products in the region and the activities associated with it'. The wine route is thus closely allied to agri-tourism but is focused on the exploration of a specific brand or collection of brands. According to Brunori and Rossi (p. 410), this exploration yields a variety of touristic experiences. The tourists

get the chance to visit a wine farm, to take part in wine tasting, purchase wine, visit a vineyard or a local museum that gives them information about the wine traditions and the history of the region. Often there is also an opportunity to stay in agri-tourist accommodation, taste the culinary specialities of the area and buy products typical of the region while enjoying the landscape.

Thus, the promotion of quality brands can segue into the promotion of a specific territory, its products, and its producers. In fact, Brunori and Rossi

(p. 411) claim that an initiative such as the wine route relies upon the generation of what they call 'structured coherence', a process that 'adds value' to individual products whether in wine, gastronomy, or accommodation. Thus, when we add together the efforts of individual producers, the consortia that govern the brand and the organizational networks that underpin wine routes, we can see how the assertion of quality through high-profile premium products can ultimately be integrated into a developmental approach that galvanizes a variety of differing actors and stakeholders in pursuit of a common goal—territorial integrity. In short, wine is a traditional Tuscan product that has been revalued in recent times and this process of re-evaluation has enhanced both the status of particular branded products and the territories in which they are produced. In short, as we shall see below, this approach shares many characteristics of the Tuscan model as it has come to be elaborated in recent years.

Institutional Development and the Emergence of the Tuscan Model

The three case studies presented above show how changes in the context of consumption have presented new opportunities for producers of traditional products. Consumers have actively begun to seek out the diverse quality foods that are seen as intrinsic to Tuscan cuisine. This cuisine has also come to be valued as one of the finest in the world. It has thus attracted affluent middle-class consumers to Tuscany in the context of a growing tourist and agri-tourist industry.

Yet, the re-evaluation of Tuscany's traditional products has not necessarily led to production practices simply becoming 'frozen' in time. While product quality is usually rigidly maintained (especially when it is regulated by legislation such as the PDO appellation), innovative changes in production, processing, and marketing are evident, as the case studies show. Thus, we might speculate that the new consumption context allows a new development path to emerge, one that might be characterized as 'innovation in line with tradition'. This path allows for the mobilization, rather than the displacement, of existing assets such as small-scale production facilities, close ties between producers and processors, and diverse local environments.

There are, however, interesting variations between the case studies. Brunello di Montalcino emerged out of scientific innovations in winemaking in the mid-nineteenth century and has more recently been subject to processes of standardization and rationalization (as the example of Villa Banfi showed). Despite the degree of change in processes of production, quality has been maintained, perhaps even enhanced, in part because a robust regulatory structure associated with the PDO legislation and the Consortium has been put in place. In the case of *lardo di Colonnata*, much less innovation has been evident yet this product is ironically threatened by inferior imitations. This threat stems directly from the producers' failure to acquire exclusive property rights through the PDO appellation, which would ideally serve to bind the product to the locality of Colonnata.

These observations highlight the importance of institutional support for individual products, even in circumstances where these products have been revalued by new consumers. Such support becomes even more significant once a territorial perspective is adopted for the 'structured coherence' that Brunori and Rossi (2000) identify as central to initiatives such as wine routes, a coherence that requires a multitude of actors to come together to generate processes of territorial integration and enhancement. Through these processes, individual products can be better safeguarded; in fact, as Santagata (2002: 15) argues, they can serve to allocate property rights and trademarks to a restricted area of production whether that be Tuscany, Montalcino, or Colonnata. In this sense, he says, 'they legally protect the cultural capital of a community localised in a given area'.

A crucial component of any territorial approach to agri-food issues is the state. As we have indicated above, for part of its history (the early post-war period), Tuscany was subject to a 'top-down' regime that sought to rationalize the agri-food sector in line with standardized models of industrial development and economic efficiency. Not only were these models inappropriate for this region but they also threatened some of the most valued assets inherited from earlier historical periods—the network of small farms, the huge variety of foodstuffs, and the diversity of local ecosystems. However, the development of an alternative, 'bottom-up' approach has been contingent on the existence of state institutions at the regional and local levels. And once again, in this regard Tuscany has been fortunate, for Italy has long held one of the most decentralized agri-food governance regimes in Europe. This has also facilitated the emergence of its distinctive approach to agri-food development.

A regional tier of government was first established in 1972 and it is notable that one of the first competences given to the regional state was agriculture. The national ministry was effectively charged with working in close relation with the new regional bodies. However, even before the creation of these bodies some regional state activity was discernible. Picchi (2002) notes that local authorities, chambers of commerce, social movements, and other such groupings were already working cooperatively within regional committees prior to the 1972 legislation. These committees served to link together policy proposals and the likely impacts on specific territories across Tuscany. As a result, 'a solid coherence between political programmes and projects and the territories involved could be constructed... the territory functioned as the "matrix" within which policy was to be discussed' (p. 206).

This territorialization of policy was taken further towards the end of the 1970s when thirty-three Intermunicipal Associations were formed. As with the regional committees, local authorities, chambers of commerce, NGOs, and others were all involved in policy formulation. These associations were focused upon supposedly homogeneous geographical areas so that policy proposals could be precisely tailored to differing spatial circumstances. Picchi (p. 208) believes that this territorialization of policy 'involved a high degree of clustering of interrelations between enterprises, a high level of cooperation between enterprises and local institutions, a common development project, and a clear identity'. We perhaps begin to see here the emergence of policy 'from below', although the Associations were short-lived—within a decade, their competences had been transferred to the provinces.

Thus, by the advent of the crisis in the Tuscan agri-food sector in the 1980s, a localistic structure of policy-making had already come into being. The provinces and the regional agencies associated with agri-food issues were therefore well able to see how local initiatives meshed with general policy proposals (whether at the regional, national, or European levels). In particular, there was recognition that some of Tuscany's most successful products—wine and olive oil, for example—were attempting to build market share on the basis of short supply chains. Such recognition was enhanced by the formation of a regional development agency with responsibility for agri-food—ARSIA—in the early 1990s. According to Brunori (2005), ARSIA focused its activities on strategic points in the broad agri-food networks by building knowledge platforms, coordinating research and development, and drawing down research funds from the EU. ARSIA has also worked to identify the large number of local products that exist throughout Tuscany and has sought to develop appropriate marketing and development strategies for these products. In some cases, this might entail applying for PDO or PGI status; in others, it might mean developing cooperative marketing relationships or strengthening links into local markets (both traditional and touristic).

By the time the Rural Development Regulation was implemented at the end of the 1990s, the Tuscan region already had a well-developed and well-coordinated institutional structure able to facilitate implementation (bolstered in the intervening period by the establishment of ARTEA, an agency with responsibility for disbursing all subsidies available under the CAP, in 1999). Once again, as Brunori (2005) points out, the construction of the RDP proceeded on the basis of what he calls 'an intense period of consultation between all stakeholders'. Through this process of consultation, the 'bottom-up' approach was reinvigorated, especially as each of Tuscany's thirty provinces and *Comunitá Montane* were to be given their own regional development plans (that is, they would be able to choose from the available menu of options what was most suitable for their requirements and would manage the bulk of financial resources).

While there was some complaint that the final Rural Development Plan put in place a rather restricted set of policy options for the provinces, nevertheless the whole process was driven by extensive local involvement across stakeholder groups. Not surprisingly, then, the plan itself strongly emphasizes the need to 'develop in ways that are compatible with values and traditions inherited from the past' (Regione Toscana, 2000: 1). It also calls for a 'multidisciplinary' and 'multisectoral' approach to rural development policy, taking into account 'environmental, cultural, economic and social parameters so as to have multiple beneficiaries'. Specific objectives include:

- to enhance the competitiveness of farms, agricultural income, and quality production by relying on farmers' capacities to improve quality (i.e. typicality and naturality) of their products, thereby obtaining higher profits;

- to conserve and enhance the environmental quality of rural areas. Agriculture can contribute to environmental quality [especially if it is focused upon quality products], which is an essential resource for tourism.

The plan goes on to outline specific measures that help meet these objectives, including efforts to raise levels of investment on farms in order to foster innovation, the promotion of training activities in order to 'facilitate a reorientation of agriculture', the provision of grants that can enhance processing and marketing initiatives and agri-environmental schemes that increase the symbiosis between agricultural production processes and valued local ecosystems.

We can see here that the RDP takes us a little closer to a distinctively Tuscan approach to agri-food development. In fact, local policy-makers are explicit about how the opportunities offered by devolution in Italy and legislation from the EU have enabled the emergence of a new approach within the region. In an interview, one highly placed administrator in Tuscany said:

I believe that the rural development policies designed and implemented in Tuscany have been very well informed by the indications provided by Agenda 2000, especially with regard to the need for an overall commitment centred not just on the basic and traditional support for farmers, but also on the conditions for the development of entire rural territories. Tuscany has made a strategic choice here. We went from an agricultural development entirely linked to the agricultural product, with all the difficulties that this implies as a consequence of bad harvests, drought, problems related to the prices and the market, it was a residual type of agriculture, to an agriculture centred around rural development. This agriculture has chosen three strengths: the quality of products, environmental sustainability and food safety. This view of rural development goes beyond agriculture: it is linked to tourism, craftsmanship, life quality, to a number of factors that turn agriculture and rural development into a value, rather than just a product. Tuscany has won the bet in this respect. In the past 'rurality' had a negative connotation, today 'rural' has a positive value, in terms of quality of products, quality of life, environmental sustainability and development of services.

He went on to say:

in short, we went from a sectoral development to a concept of territorial development by evaluating and implementing the concept of multi-functionality. Of course this is easy in a rural world like the Tuscan one, which can count on an extraordinary environmental, landscape, historic and cultural patrimony. However, today we are thinking of an agriculture that implies responsibility also from a social and environmental standpoint and that, more important, plays active and positive functions on that ground also. And we are thinking about a rural development made of many professions, many jobs, many opportunities, many figures beyond that of the traditional farmer-producer.

Thus, the 'Tuscan model' begins to come into view. It implies taking a territorial, rather than sectoral, approach and requires a holistic, rather than segmented, view of the relations among economic, social, cultural, and environmental factors. It is based on 'subsidiarity', the notion that decisions are best taken at the lowest level of the governmental hierarchy possible, and recognizes that subsidiarity is especially important in the agri-food context because food production necessarily entails the use of land, a resource that draws together the various factors mentioned above. The Tuscan approach also requires extensive consultation so that all stakeholders gain some degree of 'ownership' over the policies adopted. While such consultation can lead to long and drawn-out processes of implementation—and can also encourage more conflict in the short term—it serves to ensure that implementation runs reasonably smoothly given the commitment of so many actors to the strategy. In short, the Tuscan model emerges from below—as a matter of course it cannot be imposed 'from above'—and draws together all the territory's various assets into a strategy of territorial enhancement. Only in this way can the distinctive virtues of Tuscany's agri-food sector be effectively safeguarded for future generations of producers and consumers.

Conclusion

We began by suggesting that Tuscany comprises an Interpersonal World of food, one that is built around close links between producers and consumers. Much of the material presented above has reinforced that view. It tends to show that the diverse local products to be found throughout the Tuscan region rely upon short supply chains and direct relations between sellers and buyers. Further, the continued existence of these products is also assisted by a political structure that is both localized and inclusive. Thus, policies and strategies are precisely targeted on the needs of producers and processors in defined geographical areas. In other words, policy is oriented to the management of holistic territories rather than discrete sectors.

Yet, this Interpersonal World is not one where time stands still. First, processes of innovation are ongoing. This is most clearly evident in the wine sector, where the large returns to be made on the sale of upmarket, quality wines has encouraged extensive investment in new techniques of production and processing. To a lesser extent, innovation is taking place even in some of Tuscany's most traditional products sectors, such as, for instance, olive oil. Thus, traditions are maintained at the same time as they are modified in line with new consumer expectations. Second, the market is changing. In particular, large numbers of affluent consumers from outside the region are entering as tourists to sample local food products. This movement has been bolstered by the extensive provision of agri-tourist facilities (which also serve to give the farm household an alternative source of income, perhaps maintaining their involvement in the provision of local products). Again, change and continuity go hand in hand.

As innovations and change occur in the production and consumption spheres, Tuscany may have to face new challenges and threats. Our *lardo di Colonnata* case study, for example, reveals the existence of a tension between the localized quality of a product and the scale of global demand. Potentially, this may lead the region to face a 'conventionalization' of the agri-food sector. As Guthman (2004) argues with regard to California's organic sector, and as we will describe in Ch. 5, this is a process of appropriation of the most high-value crops and the most lucrative segments of an alternative food chain by agri-business firms. This would lead to 'agro-ecological enfeeblement' (Guthman, 2004: 310), such that the alternative sector would cease to be substantially differentiated from the conventional one. For instance, as Guthman (p. 312) suggests, if expectations of intensification become embedded in land values, the cost of land would make conventionalization hard to resist.

To deal with these threats, as we have described in this chapter, in recent years regional authorities have developed a loosely defined concept of the Tuscan model. Beyond rhetoric, this concept has provided a regional platform around which Tuscany has successfully built a significant degree of consensus among different actors over the future of its agri-food policy. As stated above, this has promoted a bottom-up approach to agri-food development which has given all stakeholders some degree of ownership over the policies implemented.

However, it is important to consider that Tuscan farmers, like all farmers, operate within a larger political economy. The region has been successful at shaping its own model and creating an institutional fabric that supports it. Yet, for this model to become sustainable, political interventions are needed at scales beyond the region. In the context of an increasingly complex regulatory structure such as the one we have described in Ch. 2, only a concerted action involving the national and supranational levels can realistically prevent threats such as the 'Californiaization' of Tuscany.

5

California: The Parallel Worlds of Rival Agri-food Paradigms

Introduction: Separation or Integration?

The aim of this chapter is to explore the nature of the contemporary agri-food worlds—the conventional and the alternative—in California. More specifically, we ask: what are the variations within each world? What sources of contestation are leading to (1) convergence and potential appropriation by the dominant agri-industrial complex; or (2) separation and real ecological modernization; or (3) a sort of coexistence and spatial multifunctionality and regulation of the two systems?

In this chapter we make some preliminary assessment of the agri-industrial pathway that distinctively marks out California as one of the most highly productivist[1] agrarian regions in the world. This region has applied successive waves of capitalist and endogenous development, with or against a series of 'obstacles'. As the literature has traditionally emphasized, the history of agri-food in California is the history of a tension within a regional brand of agrarian capitalism continually wrestling with its own contradictions between economic accumulation and social legitimacy.

The chapter first examines the historical and contemporary dynamics of the agri-industrial paradigm as it has played itself out in this bountiful but peculiar agrarian space. Specifically, it describes how the agri-food system in California has (quite successfully) attempted to overcome 'the obstacles' of what we term 'first', 'second', and 'third' natures. More so than any other region, California has developed since 1849 an agri-industrial dynamic that continues to exploit its natural and social conditions in ways that sustain an exceptional and endogenous form of 'agri-cultural economy'.

After exploiting the natural resource 'initial endowments' through a very effective 'extractive' mode (i.e. 'first nature'), the agri-industrial paradigm assembles a specific form of fictitious circulation of capital, goods, and services. This creates a 'second nature': a longstanding framework of flows

---

[1] 'Productionism' is used in this chapter to refer to the overall food system orientation which is geared to maximizing production through the setting up of regulations, production and marketing arrangements throughout the food supply chains. 'Productivism' holds a narrower connotation which refers largely to (primarily) farm-based increases in both the scale and intensity of land-based production.

of capital and labour, infrastructure and technologies, which provide a superstructure for the state to overcome the well-documented obstacles of labour and production time in the agri-food sphere.

However dominant or celebrated this peculiar model becomes at the end of the twentieth century, we see another set of profound challenges ahead. These are 'third nature' obstacles which were in part created out of the very success of a century of Californian agri-industrialism. Ranging from consumer and environmental pressure to the rising power of corporate retail capital, these concerns create a new dynamic terrain for the agri-industrial system that we analyse here by looking at the fruit and the dairy sectors.

The second part of the chapter examines the quite longstanding struggles of alternatives against the prevailing paradigm in California. At the very least, it is suggested that these represent a new 'space of articulation', one which may be less coherent, but which shows some signs of 'autonomous relocalization'. This dynamic is producing a more variegated set of producer–consumer linkages in agri-food, suggesting that there may indeed be two Californian agri-food worlds.

Seeing the Exploitative Vista

Speaking at the Californian State Agricultural Society in 1889, William H. Mills, a land agent for the Southern Pacific railroad, foresaw the significant global comparative advantages that existed for Californian agriculture (quoted by Stoll, 1998):

the competition of soils and climates is immediately present. . . . In these markets, we see the fertility of the soils and the favouring conditions of climate competing with the environment of every other portion of the world: . . . In every market there are immediately present the effects of the system of labour, the methods of production, the favouring conditions of soil and climate; they meet face to face; distance no longer divides them. Their economic presence has become equivalent of physical contiguity.

Mills claimed that California could become the 'orchard of the world' and that it was turning the von Thunean principles of distance on their head. In short, it was making a fool of distance and nature as barriers to agricultural development. By the 1930s, as Stoll (1998: 181) argues,

California fruit business represented industrial farming at its apex: the almost complete separation between farm production and consumption and the dedication of soils in a vast region to consumers far away. Though nature presented a set of ecological options making possible a great diversity, the growers' particular reading of nature led them to plant a limited number of plants in monocultural strands. Determined to enjoin California with the emerging national economy, they invested in labor practices, chemical inputs, and market-organisations intended to sustain specialised crops. People and nature served the growers in a singular capacity, but the growers refused to serve either in return.

These initial endowments set in train a course of dynamic agri-industrial development. Many writers have documented the innovative and distinctive features of this 'Californian model' (Allen, 2004; Walker, 2004), which is seen as representing the leading US agricultural region in terms of production and value, as well as being the premier home of alternative and food security movements. As such, the California model provides us with a distinctive and valuable insight into a world of food where the intensity of the 'battleground' between the alternative and the conventional is at its highest. To explore the architecture of this battleground, we will focus in particular on the power relations in the Californian agri-food system.

California as an Innovative Region: First, Second, and Third Natures

First Natures

In his fascinating agrarian history of California, Henderson (1998) admits that any student of the state faces the problem of the sheer complexity of its agricultural space. A kaleidoscope of varieties of crops has been grown, usually at productivity levels which far exceed the US national averages. Indeed, as Henderson argues, the unevenness in natural conditions and in the social and economic frameworks built up in certain places and in certain sectors stimulated capitalist accumulation through intensive agriculture.

Beginning with wheat in the gold rush era, sustained productivity increases were yielded on the back of the technological innovations, which were first and foremost based upon a unique set of 'first nature' physical endowments. Stoll describes how Californian agriculture took advantage from the start from being located 'in the rain's shadow':

the Pacific High regulates the rain, but the mountains allocate it. Storms from the ocean drop some of their moisture on the coastal plain before encountering the Coast ranges, a series of parallel ridges that run north–south, from Los Angeles to the Oregon border.... Parallel ranges traversing the state create hundreds of valleys. Much of the state's agriculture came to be conducted on these grass-covered prairies in the years after the American takeover, and the gold rush of 1849.

The great valleys, the Salina and Santa Clara south of San Francisco, the Napa and Sonoma to its north, the Orange and Los Angeles to the south, and those composing the Central Valley, which runs north–south for 450 miles, contain the river basins of the San Joaquin in the south, the Sacramento in the north, and the Delta area abutting San Francisco Bay. These arteries and valleys provide variable but rich bases for intensive agricultural development and specialization. Bounded by the Mexican deserts to the south, the Sierras and Nevada to the east, and the forests to the north, California came quickly to represent a sort of agricultural island, distanced from the rest of the US in terms of markets but with its own geophysical features which allowed for commodity specialization across different climatic zones (Bill Friedland, personal communication, 15 March 2005).

These first natures became the basis for capitalist agricultural development, which intensified production and raised productivity in California well beyond other regions in the US. By 1930, California was the greatest fruit-growing region, contributing between 60 and 100 per cent of the US production of table and raisin grapes, apricots, prunes, lemons, figs, almonds, and walnuts. By 1955, the average yield of tomatoes and cotton was twice the national average; milk per cow was ahead of all other states; and by 1980 strawberry yields were five times the national average.

First-nature natural resource exploitation was, then, central as a starter for agrarian capitalist development in California, and it was integrated with the exploitation of minerals and forest lands and the parallel developments of urbanization. Following Cronin's (1991) classic account of how Chicago became the centre of regional commodity circulation in the Midwest, profiting off the circulation of wheat, lumber, and meats from the surrounding rural areas of Wisconsin and Illinois, Walker (2001) documents the peculiar, but dramatic evolution of 'Californian capitalism' as it is based upon intensive but variable forms of resource exploitation. It was an exploitation of first natures, and one that sustained itself by creating the superstructure for a second nature.

Second Natures: Circulating Capital, Commodities, and Technologies

While the natural advantages and bounded geographies of California may have first stimulated its peculiar path of agrarian capitalism, it has been the dynamic social development of its organising forms, its private property rights, its generalized and liberalized market structures, its wage labour arrangements, and its flows of finance and money capital between the urban and the rural that instituted a framework within which such resource endowments could be further capitalized. Once the chief obstacle to the imposition of white individual farm occupancy had been removed (with the half a million indigenous residents reduced to 10,000 within a century), a 'free' system of labour (which involved large imports of migrant Mexican labour) and competitive markets could unfold quickly. Small settler farmers proliferated from the 1880s to the 1920s as extensive arable and grazing lands were broken up for more intensive systems of fruiticulture and dairying (Liebman, 1983). These macro trends hide the development of social struggles between extensive and irrigated lands and a variety of ownership and labour patterns that emerged in different parts of the state.

By 1925, there were 136,000 farms and many of their occupants were of urban origin, innovators and experimenters who had strong links to urban finance houses. While the rural spaces were to be the domain of the petit bourgeois family farmers, their wealth was banked and circulated back into reinvestments in agri-business and capital stocks associated with the burgeoning financial centres of San Francisco and Los Angeles. In short, an agri-industrial complex was born. Karl Marx, writing in 1880, recognized the importance of the rapid centralization of capital taking place in California (quoted by Walker, 2001: 1900): 'California's regional capitalism was a mighty engine of resource discovery, extraction, cultivation and plunder that left no stone unturned in its efforts to wrest the maximum reward from the land.' Walker sees this as a 'pure' form of capitalist development that not only held the three key features outlined by Marx—private property controlled by a capitalist class, the exploitation of wage labour, and monetary investment for profit—but is also qualified by three distinctive regulatory features or infrastructures. These include: (1) the expansive and expansionist notion of agrarian commodity systems (and their attendant social division of labour); (2) the vital relation and transformative effect on nature in production and commodity circulation; and (3) the distinctive social organization of production and the business management side of the industry. From the 1850s, California became the first and most complete example of industrialized agriculture (Jelinek, 1982), which promoted an agri-industrial complex based on a hierarchical and diverse division of labour, from the farm to the factory. This created an 'integrated business system' which involved the flow of materials through commodity chains (Friedland, Barton, and Thomas, 1981), the interaction of different elements of the agri-industrial complex and the organization according to modern business practices.

Another distinctive feature was the lack of social resistance to the onset of this new agrarian capitalism. Unlike other regions of the US, such as for example the Midwest or the East Coast, where settler agricultures based upon family farming preceded the development of agri-business capitalism (see Friedmann and McMichael, 1989; Guthman 2003), California lacked the history and sunk costs of 'pre-capitalist' farming communities. Its 'island status', initially at least, reduced large-scale social resistance to its agricultural revolution.

Even though, as we shall see below, alternative visions and 'paradises' of small-scale agriculture, such as small horticultural enterprises at the end of the nineteenth century, the New Deal of the 1930s, and the organic movement of the 1990s, periodically emerged, these movements are all compromised into variants of an agri-food complex built upon a super-productionist paradigm. In this sense, California represents a quintessential exemplar of super-productionism, whereby, as we discussed in Ch. 3, the production sector is designed to produce more and more and the processing and retailing sectors tend to design and sell more and more. In California, this led to the 'redesign' of plants and animals and the scientific reconfiguration of first natures' outputs.

For Henderson (1998), this capitalist agrarian experience (of *second nature*) in California turns much of the principle of the widely recognized distinctiveness of agrarian capitalist development on its head. Following Kautsky (1988), as we explained in Ch. 3, the Mann and Dickenson thesis (1978) has long explained the persistence of relatively non-capitalist production forms, such as family labour and individualized property rights, as a function of capital's inability to reduce the gaps between 'production time' and 'labour time' in agriculture. The simultaneous rise of regional finance capital and agricultural capital in California suggests that disparities between labour time and production time can in fact become new opportunities for capitalist development. According to Henderson (1998: 32, emphasis added):

this same nature-centred production poses opportunities for capital precisely because [capital] must circulate and precisely because the disunities of production and working time (necessitated by natural processes) and capital's time in circulation (in part, nature as distance or as space) exist. That is, if these things exist for potential capitalists *as a cost to be averted*, then they exist as an *investment for capitalists* looking to fund anyone who does get involved in having to cover the cost.

It is this integration among finance, agricultural, and agri-business capital that lies behind the sensational growth of Californian agriculture. The geography of credit takes on a special importance here. Its role was crucial in transforming many farmers into what Henderson calls 'capitalist–labourers' who functioned at times as a deployer of capital and employer of labour and at other times as a more or less proletarianized labour source for the owners of credit. What becomes significant in the Californian case is the degree and type of capital circulation and whether that circulating capital confronts a 'capitalist–labourer' farmer or a migrant labourer. In this sense, capital again confronts (second) nature, that is, nature in the form of the human body. As Henderson (p. 41) argues: 'workers are sets of biological processes and energy flows for which capital has only partial substitutions (robotics). They are themselves obstacles to capitalism. Bodies persist. That they are waged bodies is a capitalist solution. That they are waged bodies is a capitalist problem.'

For Henderson (1998), there are always partial solutions to the problems thrown up by capitalist circulation. For instance, the specific circulation times of capital and credit, work as human and mechanical labour, and the vagaries of nature give rise to the specific and variegated geography of Californian agricultural production. As Mitchell (2000: 474) summarizes:

patterns on the ground do not autonomously give rise to other patterns, rather, complexly intersecting patterns of circulating money, bodies and nature create new patterns—new obstacles and new opportunities ... only overcome or realised as the result of never-ending struggles as to who is going to control the point of production, the point of credit, and the point of labour reproduction.... Human agency sits right at the centre of Henderson's theory of regional development: the decisions of innumerable actors as they interact with shifts in capital, processes of nature, and unruly bodies continually mold the logic of capital, commodity and labour circulation.

For example, in the early 1920s San Francisco's Anglo and London National Bank became a broker and promoter of irrigation district bonds. As Henderson (1998: 122) explains, 'the borrowed money becomes a part of the [irrigation] industry's productive capital...the installation of an irrigation system is a capital investment such an industry might make—and results directly in the creation and addition of new wealth to the security back of the debt'. For Henderson (1998) this represents an example of the 'geography of fictitious capitals over supra-local space'. In the process, the meaning of nature changes:

nature has in a sense led the district's farmers to over-accumulate periods of slow turnover—therein the investors' opportunity. But it is not so much the direct transformation of nature that constitutes the opportunity. Rather they rely upon a broadly differentiated space of 'second nature', the geography of human-produced differential rents—sites of different capitals in different locations, circulating along varied timelines and producing different 'needs' and different 'yields'.

Fresno County, for instance, from the early days functioned as a rural centre of production within a wider matrix of flows of commodities, labour, and capital. In the change from an arable and beef-producing area to one of intensive fruit cropping, finance capital was a key agent. When farmers needed to obtain credit to expand production or to manage the disunities between production investments and working and growing cycles, they turned to the (often, grain-based) financiers in the San Joaquin Valley or in San Francisco. The latter was the main financial centre, a regulatory centre and a transportation centre for the grain trade. Capital mobility in agriculture was thus highly developed by the start of the twentieth century, with rural banks keeping deposits in San Francisco and the gold and silver reserves being quickly translated into cash for investment and credit bonds.

This regionally mobile development of urban and rural-based credit fuelled the productivism of Californian farming and stimulated the development of mechanical technologies and the capitalization of the land, not least through irrigation technologies. Privatized forms of credit and finance capital became the main 'regulator' of the countryside and encouraged greater and greater local and regional specialization. Finance capitalized the land, forcing producers to get as much production as possible from it. This major dimension of regional development throughout the twentieth century also facilitated the growth of agri-business firms.

The state government played a supportive role for capital and agri-business by developing educational, research, and extension programmes to enhance production and make innovations. This publicly funded knowledge creation was rapidly applied to commodity specialization, especially after the growth of the extension service in 1914, and it also fostered cooperative arrangements among 'growers'. For instance, the Commission Marketing Act of 1915 and the Fruit Standards Act of 1927 provided state control over quality in accordance with the co-op model, and the latter empowered grower-run marketing boards for every major commodity. The dairy sector also had its own State Dairy Bureau and a Pure Milk Act so as to assure quality and to limit competition.

The development of a distinctive and endogenous agrarian capitalism in California was thus based upon collaboration among industrial and finance capital, the regional state, and producer groups. These relationships were not always harmonious (especially concerning labour rights) but they continued and sustained an innovative culture. California became, Walker argues, a *learning region par excellence*. Innovative machinery used in one location was quickly followed by large amounts of capital for development and marketing there and elsewhere. Overall, this was an essentially endogenous economic development trajectory, based upon Californian banks, builders, and businessmen.

This agrarian capitalist framework was essentially built upon 'a mini-state within a nation-state'. In fact, while California busily assembled and developed hybrid forms of private and public regulatory structures of its own to facilitate its super-productivism, it tended to shun, from the start, the intervention from the federal state at large. As Walker concludes (2001: 191), this was a regionalized form of neo-liberalism that engrained itself through the course of the state's development in the twentieth century:

industry grew and continually innovated, thanks to the creative genius of skilled labour backed by lots of money and robust regional markets. The state gave capitalist profligacy a free hand, periodically reformed its grossest excesses, then stepped back to give business a free hand once again. All along the way, California's resource economy walked forward on two legs: natural wealth and social production, industry and extraction, big business and small property, city and country, state and private enterprise, capital and skilled labour (not to mention highly exploited labour), safe bets and wild speculation.

The development of second nature agri-industrialism was built upon managing and manipulating the sets of first nature initial endowments. Probably more so than in any other agricultural region, the natural *obstacles* of capitalist agriculture were indeed, for a time at least, overturned or ploughed under into further opportunities. This trajectory was built upon a particular coalition and collaborative set of relationships between production-based rural concerns and urban-based industrial and finance interests.

However, by the 1970s and 1980s, these conditions and coalitions began to confront a new set of socially and naturally constructed '*obstacles*'. These are part of a third nature and represent both a reaction to the extreme forms of agri-industrialism, on the one hand, and a growing public concern for *con-*

sumerization, rather than productionism, on the other. Moreover, they begin to cross-cut the previous harmonious integration between the urban and rural sets of interest even in the most productivist rural spaces of California. Simultaneously, the rising power of corporate retailing tends at least to match the traditional hegemonies of finance and industrial capital. In this more complex third nature realm, then, extreme productivism begins to meet a new corporate and public consumerism. In contextualizing this collision within the conventional agri-food sector, we will examine in more detail two commodity sectors: strawberries and dairying.

Third Natures: Emerging Public and Consumer Contestations

In the realm of *horticulture*, strawberry production well illustrates the contemporary phase of Californian agri-food development. In 1946, Californian strawberries represented only 6 per cent of US production; by 1988, the region accounted for 74 per cent of national production. The main reason behind this was the exceptional increase in yields, which rose from 3.7 tons to a staggering 24.2 tons per acre between 1946 and 1988—almost five times the tonnage produced in other parts of the country.

Most of the production is concentrated in a 20-mile-wide strip of land running along the central and southern coasts of California. Wells's study (1998) describes the high degree of labour 'flexibility' and exploitation that has characterized the Californian strawberry sector and its ability, through concentrated market power in the hands of berry producers, to develop a high degree of protection from the instabilities of the wider national and international markets. She argues that the organization of the industry reinforced the economic viability of relatively small producers operating in the same region and that control over hired harvest labour, in particular, became a key feature in ensuring profitability. The labour organization of the Californian strawberry sector displays significant amounts of flexibility concerning employer–employee relationships. Wells discovers, for instance, that before the Second World War a dominant form of organization was sharecropping (or share-farming), and that this almost disappeared afterwards, to return in the 1960s and then decline again during the 1970s. These were variable systems of labour control by growers that tended to minimize workers' rights, creating significant tensions and labour conflict in the fruit sectors.

The strawberry plants have been bred continually for over a century. Present-day varieties are intensely overbred (Friedland, 1998), and this implies the destruction of all other forms of life in the soil in which the plants are grown by covering the fields with plastic sheeting and injecting methyl bromide (MeBr) into the soil. This practice, which kills weeds and pests but also creates environmental problems, is currently the focus of an intense conflict between environmental groups and the strawberry growers, who need to use it in order to keep production costs down. Indeed, growers continually face an intense cost-price squeeze: overall input costs continue to rise for conventional strawberry producers (more that $20,000 per acre) and 98 per cent of input energy comes from non-renewable sources. To secure a return on this investment, producers must rely upon pre-plant fumigation with methyl bromide, plastic mulch, drip irrigation, pre-plant chilling, fertilization with slow-release nutrients, foliar applications of synthetic pesticides, and concentrated semi-permanent hand-labour throughout the growing season. Despite these increasing (and often hidden) input costs, producers can expect profits of over $6,000 per acre (1994 prices).

For Wells (1998) this super-intensive system is maintained as a social and moral economy based upon particularly exploitative sets of labour relations that tend to shape the social constellation of particular sub-regions of strawberry production. For instance, Salina's valley growers are predominantly Anglo and farm large acreages; Pajaro growers are mostly of Japanese origin and work middle-size farms; North Monterey growers are mostly of Mexican origin and farm the smallest units. These differences, based on histories of local Anglo development and Japanese and Mexican insertions into the labour and property markets, have important implications for the differential social organization of production. Anglos, working the largest farms, tend to have hierarchical management structures, whereas the Japanese and especially Mexican farms have closer, informal relationships with their workers. This particular international mixing of productive and labour forms, combined with the mobilization of scientific efforts to continue to breed the 'super-strawberry', provides what seems to be a continually profit-making sector. However, there are also significant labour and environmental instabilities, which so far have been tackled on a short- or medium-term basis. In fact, despite a long history of worker resistance and political mobilization, union membership continues to decline, contracted labour is rising, there is more mixing of ethnic (especially Mexican) labour, and the significance of labour legislation is decreasing (Wells and Villarejo, 2004).

Despite the continued predominance of Californian strawberry production nationally and regionally, the industry faces significant vulnerabilities associated with its environmental impacts and its increasing dependence upon corporate retailing buyer power. As we will discuss in the next section, these trends suggest that the conventional systems of strawberry production will continue to be a growing source of social conflict in the region.

Third Nature Hits Back: The Onset of Methyl Bromide Regulations

In January 2001, the California Department of Pesticide Regulation (DPR) introduced controversial MeBr application regulations, which had a significant impact on strawberry agriculture. Further and stricter regulations were prepared at the end of 2004 as a result of local and regional public concern but also as part of the Montreal Protocols proposed ban.[1] The DPR state-level regulations aimed at reducing human exposure to MeBr. For each fumigation site, the DPR regulations stipulated dual buffer zones where MeBr could not be applied to the soil. The buffer zones depended on the application rates (i.e. pounds of MeBr per acre, method of application, and the proximity of schools, houses, and other occupied buildings). The regulations also contained worker-hour restrictions and required growers to notify neighbouring residents when they were going to use the chemical.

Economic analyses of the effects of these public regulations suggest that they are significantly affecting the industry (Carter, Chalfant, and Goodhue, 2002; Carter et al. 2005). The main economic impacts include: forgone profits from sales of processing berries due to a reduction in season length; added labour costs due to the longer fumigation periods; loss of land for intensive production associated with the buffer zones; and public notification costs. The regulations came under close scrutiny in the courts as farmers fought to have them neutered. The regulations, it is claimed, forced some smaller growers out of business or obliged them to use alternative and less efficient fumigation procedures, which were seen as less effective at controlling pests.

By 2005, the use of MeBr was still not completely banned and the controversies over its effects continue. Carter et al. (2005) show that MeBr applications did not substantially decline between 1996 and 2003 and that the relative share of MeBr applications in relation to other crops has actually increased. Producers' organizations (such as the Strawberry Commission), however, continue to focus on the negative effects of this partial legislation, pointing to the impact this has on the industry, the decline in efficient production and the rise in imports from countries, such as Mexico and China, that are not legislating against its use. Alternative treatments, such as Telone and 1,3-D, are also seen to be harmful to humans and generally record lower yields. The chief of the Strawberry Commission believes that there are several well-entrenched myths associated with MeBr (Jones and Prescott, 2005). It is estimated that farmers' yields would decline by 15–20 per cent if the chemical were banned completely.

A technological 'solution' potentially lies in genetic engineering. Researchers in private firms and universities in California lead the way in developing biotechnology in strawberries (Whirty, 2000). An Oakland-based company announced that it had grown strawberries that were resistant to the herbicide glyphosate, commonly known as RoundUp. Company representatives argued that their ability to induce a tolerance to glyphosate would allow strawberries to survive sprayings of RoundUp and that this could be used as a substitute for MeBr within a few years.

However, the main barriers are considered to be consumer reaction and legal issues regarding implementation. As one of the key scientists argued,

---

[1] According to the EPA, US farmers purchased 38 per cent of the global MeBr in 1996.

competitive regions such as New York State cannot simply copy the new GM developments because the technologies are now owned by the companies themselves. Moreover, as he states (quoted in Whirty, 2000): 'all of these modified products have benefited the companies and the farmers, but there's been no benefit to the consumer at all.... Monsanto sells more herbicide, farmers have an easier time in dealing with weeds and can cut their costs, but the consumer doesn't get anything out of it. I can understand why there is a backlash.' In California, specifically, the growing ex-urban populations who have suburbanized the deeper rural areas are expressing the most public concern. As Thacker (2005) stated:

weeds are less of a problem [in California, where the fungal infections usually kill crops]. There, farmers do not need the added herbicide. However, California farmers compete with developers for land, and many fields border houses and buildings. EPA considers Telone a probable human carcinogen, with moderate toxicity to wildlife. Application also requires a wide buffer zone if the field abuts an occupied structure.

With regard to the rise of *fruits and vegetables*, it is estimated that 55 per cent of the total value of Californian agriculture ($26 billion) is provided by the fruit, vegetable, and nut industries. As a result of their predominant market share, Californian producers and processors have traditionally held unique opportunities to exercise control over the markets for those commodities, and this has been supported by specific state policies for marketing, grower, and cooperative arrangements. However, over recent decades, as elsewhere, there has been an increasing marketing bill placed on growers, which repre- sents a shift in the appropriation of value towards the retailers. The farm share of the value of the market 'basket' (i.e. the average quantities of food coming from farms and purchased for consumption in the home), which remained stable at 40 per cent between 1960 and 1980, has declined rapidly since then—to 30 per cent in 1990 and 21 per cent in 2001. Farm values have traditionally accounted for more than 50 per cent of retail value for animal products such as meat, dairy, poultry, and eggs, but these shares have now fallen to below half. The farm share for fruits and vegetables tends to be much lower and does not vary much between processed and fresh products (Carman, Cook, and Sexton, 2004). With more Americans spending a higher proportion of their incomes both in the main concentrated retail sector and in restaurants, farmers' share of the total retail value for the major food commodities was down to 19 per cent in 2001, compared with 41 per cent in 1950 and 24 per cent in 1990. For fresh fruit the farm share of retail value is even lower (16 per cent for fresh fruit and 19 per cent for fresh vegetables, 2001), falling from 26 per cent in 1980.

These trends are affecting Californian production systems with regard to the relative amounts of value the fruit growers are able to capture from the sale of bulk, conventional goods— in which they have long had comparative advantage. This is linked directly, as explained in Ch. 3, to the recent consolidation of the US corporate retailer sector (Wrigley, 2002). In 2002, retail chains (defined as a retailer operating eleven or more stores) accounted for 83 per cent of supermarket sales, compared with 54 per cent in 1954.

For instance, the arrival of a fast-developing Wal-Mart is driving non-value-adding costs out of the food supply system, raising competitive benchmarks for other retailing outlets. Wal-Mart opens over 200 new supercentres per year in the US; by mid-2003, it owned 1,333 supercentres in the US that sold Californian fresh fruits and vegetables. Increased retailer buying power is restructuring the traditional fresh fruit and vegetable markets, creating 'preferred supplier' contracts and intensifying competition among suppliers for shelf space. While fresh vegetable and fruit consumption continues to rise (15 per cent in the US between 1976 and 2002), so does product differentiation, with fresh-cut fruit leading the increasing demand. The amount of fresh produce in US supermarkets has expanded dramatically. It increased from an average 133 items in 1981 to 350 in 2001, reflecting a growing diversity of consumption practices and more demand for speciality and ethnic fruit and vegetables, as well as the growth in the diversity of fresh-cut, value-added, and convenience products.

In California, direct price and income supports apply to only a few major crops, such as rice, cotton, and dairy. The role of the state and federal government in the mandatory marketing programmes is mainly that of a facilitator. According to Carman, Cook, and Sexton (2004: 117), 'government provides the legal framework for industries to take collective action, but decisions on whether and how to use these programmes are made by the industries, and they are self-funded'. Today, perishable crops that need to be harvested, sold, and marketed within a short time-frame tend to give growers declining amounts of bargaining power in dealings with buyers, while the consolidation of purchasing within the hands of a few larger buyers (often operating for the corporate retailers) raises growing concerns about the oligopsony exploitation of producers.

As stated above, Henderson (1998) highlighted the opportunistic role of financial capital in exploiting the distinctive disparities between production and labour time in Californian agriculture. The new trends in the marketing of fruits and vegetables now suggest a similar opportunity for a more consolidated and consumer-driven corporate retail capital, which increasingly sells Californian fruits and vegetables across the country, but it does so by extracting more value from the producers.

We see then that despite the continued predominance of a vibrant and intensive fruit sector in California, based upon expanding and more diversified markets, the gradual consumerisation of agri-food—in the form of both more public environmental concern over the potential harmful effects of its technologies on the one hand, and of the rapidly consolidating buyer power of the retailers on the other—is beginning to shape and constrain the sector. In short, while much the sector still relies upon its peculiar brand of

Californian resource 'flexibility', it also has to engage with a more contentious and competitive social context.

Dairying: Innovation, Relocation, and Quality

The states of the Midwest, and especially Wisconsin, are traditionally regarded as the 'dairy states'. Even by 1970 Wisconsin dairy farmers were producing double the amount of milk compared with their western US counterparts. By 1993, however, California surpassed Wisconsin, and by 2000 it was the largest milk producer in the US, with 20 per cent of the national total. Like other sectors in the state, the success and vulnerabilities of dairying stem as much from its relationships with its cosmopolitan roots as with its agrarian ones.

If not in California as a whole then in LA county, dairying between 1925 and 1965 became a leading national player (Gilbert and Wehr, 2003). People and milk went together. This was one of the country's most populated and fastest-growing urban areas, which doubled its residential population and its dairy cows every couple of decades or so. After the Second World War, the county held one of the largest cow markets on the globe.

To understand the social dynamic behind this growth, it must be considered that the dairy farmers of LA county developed a new model that would be copied elsewhere in the US and beyond: 'dry-lot' dairying. Essentially, they concentrated cows on small plots, purchased, rather than grew, all the feeds, and then fed it to the animals on a 'zero-grazed' basis. By quickly becoming industrialized, with large herds, innovative technologies, and a heavy reliance on hired, rather than family, labour, this model created 'milk factories' and 'dairy cities' such as Dairy Valley, which contained 3,505 people and 85,000 cows in five square miles by 1960, Dairy Land, which had 600 people and 11,000 cows, and Cypress, which included 1,700 residents and 13,500 cows.

Gilbert and Wehr (2003: 484) describe an archetype whose principles were to transcend much of the industrialized dairying world:

the three incorporated dairy cities were zoned exclusively for heavy agriculture. By stabilising the land market, the state's protective zoning kept property taxes low. It also insured the ability to improve and expand the dairies without fear of complaints from non-farm neighbours. Since they were essentially composed of farms, the single purpose cities minimised municipal services such as paved roads and street lights. In effect these were agricultural areas in the midst of one of the largest and fastest growing metropolises.

It is clear that the world's first industrialized dairies trace their origins to the urbanization of LA county, where a vibrant non-agricultural land market stimulated relocation and further capitalization of intensive production. Urbanization, suburbanization, and agri-industrialization of the dairying sector went hand-in-hand and began to lead to a 'structural separation' that marked out a distinctive Californian path for dairy industrialization.

There are significant regulatory features of this agri-industrialism. As the only major milk-producing state that lies outside the US Federal Milk Marketing Order, California administers its own milk pricing and pooling rules; and, as Butler and Wolf (2000) argue, this includes a quota that, according to many competitor states, artificially and unfairly increases the prices for Californian dairy producers. For instance, in a series of Congressional hearings relating to the regionalism of the US dairying sector, Northeast and Upper Midwest legislators were 'shocked and dismayed' at what they saw as the flouting of the rules relating to production quotas. The management of the Californian dairy state order system was regarded as 'inequitable'. California's endogenous economic regionalism, initially based upon its geographical (first nature) isolation to the west of the Sierras, was once again to the fore, with its state policy seen as 'cushioning market shocks' and uncompetitive by many of those outside its boundaries. DuPuis (2002) reinforces the significance of this regulatory distinctiveness, arguing that the method of pricing milk in California historically strengthened the protective boundary between market and manufactured products by strongly restricting entry into the fluid liquid market. The state had greater autonomy in setting prices and managing markets, and this protected Californian dairy farmers from US price competition.

Competitors from the Midwest and the East had good reason to be concerned. From 1950 to 1998 overall US milk production increased by 35 per cent as a result of increases both in population and income. This was complemented by a 58 per cent decrease in the number of cows and a 223 per cent increase in the milk per cow. In California these trends were even more pronounced; milk production increased by 361 per cent and milk cows increased by 82 per cent. By 1997 the average herd size in California was a staggering 530, compared with 59 in Wisconsin and 78 nationally. As with other farming activities, this represented a significant type of regional 'structural divergence' (Gilbert and Akor, 1988), which can only be explained in terms of the distinctive combination of regionally constructed conditions.

Among these regional conditions, the distinctive state regulatory context is especially important. The Californian state developed its own milk marketing orders in the 1930s, giving itself the autonomy to experiment without having to coordinate with neighbouring states. The state can set its own pricing orders using its own formulae, and since 1969 it has set its own milk quota rules, which are geared to maintain high revenue for those producers who historically marketed in the higher-valued fluid market. This has production implications, with Californian farmers receiving a non-quota price as their marginal price, while those under federal rules elsewhere have identical average and marginal price (Butler and Wolf, 2000; Sumner and Wolf, 1996).

In addition, allowances and transfers of funds to processors from producers have been arranged differently and there are distinctive quality standards for fats and solids content of fluid dairy products. In fact, specific tests have shown that consumers tend to prefer the richer, more consistent fluid milk resulting from the California standards. For instance, minimum standards of calorie content are significantly higher for whole and low-fat milk than federal minimum standards. These have emerged out of negotiated compromises between producers and processors that led to paying producers for both fat and solids-not-fat contents at consistent levels.

While contributing to the 'structural divergence' and distinctiveness of Californian dairying, these distinctive regulatory features also expose a culture of independent negotiation among different parts of the industry within the region which has been conducive to the further intensification of the industry. Furthermore, these regulatory features need to be seen in combination with other factors, such as investment timing, technological innovation, and the increasing suburbanization of the growing Californian population during the 1970s to 1990s. It is worth considering these factors in more detail because they characterize the contemporary dynamic of the dairy sector in California.

While dairying in other states was more longstanding, it also displayed more fixity in its sunk costs. As a result, the continued use of the traditional Stanchion Barn in the Midwest and the Northeast was difficult to shift. Many new entrants in California, however, were able to make significant technological 'leaps' because of their lack of sunk costs in traditional equipment and buildings. Larger and more mechanized dairy parlours and free-stall barns gave California dairy producers an adopter advantage on Cochrane's notion of the 'technological treadmill' in US agriculture (Cochrane, 1993).

With one in ten Americans living in California today, and population growth at twice the national average, more regional milk is needed. On the one hand, this has stimulated growth in the processing and production sectors; on the other hand, rapid suburbanization has created land equity benefits for many contiguous dairy farmers. These can roll over their capital assets by selling some or all of their land and relocating and, at the same time, can technologically upgrade their operations.

This equity roll-over process stimulated reinvestment in the industry, especially around Los Angeles, in Chino County. This has given dairying in California a spatially dynamic feature, whereby encroaching suburbanization has stimulated more intensive investment and larger-scale enterprises. By the 1970s, dry-lot feeding had spread northwards into the Central Valley and around the towns of Fresno, Kings, and Kern. Once again we see the city, now as a vast suburbanizing 'frontier', providing a major stimulus for the geographical shift and technological development of a key agrarian sector. Such a shift has created much competitive regionalism in policy debates at the federal level; as Butler and Wolf (2000: 160) conclude:

as long as California remains outside the federal orders, and as long as there are regional differences in fluid differentials, commodity production and other factors that differentiate one region from another, there will be regionalism in public policy debates. The California policies have had a role in demonstrating alternative methods that later have been widely adopted by federal policy makers, as in the case of the most recent federal Milk Marketing Order Reform.

Third Nature Concerns in Dairying: rBGH and NIMB

These engrained processes of agri-industrialism are challenged, however, as we found in the fruiticulture and vegetable sectors, by third nature public and environmental concerns. Both are related to the processes described above, especially suburbanization and the growing public concerns with the continuance of the 'technological treadmill' in dairying. In particular, the development and adoption of rBGH (Recombinant Bovine Growth Hormone) to enhance milk production and the threats this poses for the loss of many smaller and medium-sized dairy farms have raised both producer and wider public concerns about the continued capitalization of the dairy sector in California (DuPuis, 2002).

The onset of rBGH stimulated the rapid growth of the organic dairy industry in the state during the 1990s. DuPuis (2000) sees this not as simply another economically productivist trend or as a reaction to the potential health concerns of using such growth hormones; rather, to her this represents the growth of 'Not in my body' (NIMB) politics of metabolic refusal. While this may not be a coherent social movement of politically conscious consumers, it is influencing the industry through a process of suburban 'reflexive consumption'. At the very least, it is stimulating smaller organic and local producers who begin to create an alternative approach to productivist dairying. For instance, DuPuis refers to the example of Straus Family Creamery, a small organic dairy company in Marshall which markets its products by conveying the need to protect and maintain family farming. Their labels state that 'dairy farms are disappearing at a rate of 5 per cent a year' and that 'going organic gave our family the chance to continue farming'.

These trends are not only associated with the rise of organics. In a study of the dairying sector in the North Bay area, Guthey, Gwin, and Fairfax (2003) contrast the high levels of spatial mobility of dairy farms and the associated intensification of production which characterize the area around Los Angeles and Central Valley with the strong suburban commitments for land conservation and environmental value. As a result of such commitments, dairying in the North Bay area is becoming more quality-based.

These counter-movements, which we will analyse in the second part of the chapter, have partly emerged out of the internal dynamics of the agri-industrial model of dairying itself. Recombinant Bovine Growth Hormone represents, perhaps, a technology too radical for many increasingly reflexive consumers. As the conventional dairy sector becomes more concentrated, so do the attentions of increasing ex-urban populations on rural landscapes that are now bereft of grazing cattle.

In short, as with the fruit and vegetable sectors, the conventional dairy sector is now confronting a new set of third-nature constraints. The question is: will this result in a further transformation of the conventional sector or, rather, in a new form of agri-food dualism, whereby an alternative archipelago emerges within a wider spatial dynamic of technologically driven agri-industrialism?

Final Transformations of the Productivist Paradigm?

As we have explained above, both the fruit and the dairy sectors in the state have developed regionally distinctive worlds of food through an amalgam of social, political, economic, and technological means. This has been, first and foremost, an endogenous process, where timing and investments and geographies have come together to create a super-productivist agri-food culture. Moreover, it seems that it has been the cultural, geographical, and regulatory distance and isolationism from the rest of the US which has helped to develop this particular and peculiar brand of agrarian capitalism and complex world of food, which is still unfolding and mutating.

Our analysis also demonstrates that there has been a distinctive role for the Californian state as well as for national regulation in helping to fuel the dynamism of the conventional agri-food sector. By 2002, despite a profound period of neo-liberal policy and what Walker (1999) calls 'governmental rigor mortis' following Proposition 13, the agricultural sector had benefited, albeit differentially, from a wide range of state supports. Sumner and Brunke (2004) estimate that these represent 10.8 per cent of the total value of output and payments (i.e. the Producer Support Estimate, which measures all direct and indirect public transfers to the sector). This is significantly less than in the US as a whole (21 per cent) and the average across all OECD countries (31 per cent). There are substantial variations across commodity groups, with rice, sugar beet, wheat, cotton, and dairy claiming the most support, with hardly any support for fruits and vegetables. More that 51 per cent of all support in Californian agriculture goes to the dairy sector; and the largest portion of this (41 per cent) goes into maintaining import barriers. Even though NAFTA eradicated trade barriers with Mexico, this was of little threat to Californian producers; in fact, import barriers have so far persisted in the dairy sector. This continues to stimulate oversupply and higher than world market price in dairy products, and subsidized exports, along with donations to domestic food programmes and international food aid. These have traditionally operated as a Californian 'surplus-vent' (see Ch. 3) for overproduced dairy goods.

Of importance too, and separate from the calculations above, is the long-standing government support for irrigated water. Much of the reservoir and distribution systems was developed by the federal and state governments. For instance, over half the water available in the San Joaquin Valley comes from major state water projects. Subsidy rates vary from $10 to $40 per acre-foot, depending on the region. Total water subsidy for California is estimated at $88 million (Sumner and Brunke, 2004), while marketing and extension services provided through the Department of Food and Agriculture cost $60 million per year.

If Marx found the Californian case a distinctive process of accumulation in the late nineteenth century, Friedland (2002) too, a century later, demonstrates how new agrarian developments and dualisms in the state tend to play havoc with our conventional agri-food conceptual categories. As he explains: 'Growing grapes and making wine have historically been considered agriculture, but classifying the current manufacture of wine as "agriculture" represents a considerable stretch of the imagination, because massive wineries resembling petroleum tank farms have emerged. Such "farming" has been expanded beyond recognition by the sales of T-shirts, wine paraphernalia, and books' (p. 364). At the same time, however, private property rights as a basis for productionism are challenged by a new consumerism (p. 367):

the alteration of landscape from oak-and-meadow rolling hills to flat open fields with vines, once accepted and encouraged to inhibit housing development, has now entered a new phase. Will preservation of the landscape (other than wilderness) preempt landowner's historic rights to do what they wish with their property? This right has been eroded in many ways, but landscape as an element of control is new, at least in this country.

In short, even though in the past the agri-industrial system was highly successful in overcoming the first and second nature constraints to which writers such as Stoll and Henderson allude, it is now confronting some new social and natural constraints that we call here elements of third nature. These represent an amalgam of consumer as well as production concerns and emanate out of the continual application (and increasingly recognized limits) of the agri-industrial technology paradigm on the one hand and of the profound, but yet not fully understood, reconstitution of relationships between the urban and the rural on the other. As after the gold rush, the fate of agri-food in California is bound up with its webs of relations with urban as well as rural cultures. Is it in the context of these new interactions that a potentially new and alternative agri-food paradigm could develop?

The Emergence of an Alternative World of Food

Does the concentration of production in the hands of fewer and fewer big operators really serve the ends of cleanliness and health? Or does it make easier and more lucrative the possibility of collusion between irresponsible producers and corrupt inspectors? In so strenuously and expensively protecting food from contamination of germs, how much have we increased the possibility of its contamination by anti-biotics, preservatives, and various industrial poisons?... And finally what do we do to our people, our communities, our economy, and our political system when we allow our necessities to be produced by a centralized system of large operators, dependent upon expensive technology, and regulated by expensive bureaucracy?... This tech-nology, in addition to so-called miracles, produces economic and political conse-quences that are not favourable to democracy. (Berry, 1981: 102)

Critical debates about the future of the American countryside and the role of agriculture in it have been traditionally marginalized by the process of agri-industrialism described above. The public and increasingly private funded research establishment has been firmly centred upon promoting technological solutions to the problems of increasing yields and reducing labour and production costs on farms. California was not allowed to develop university departments of rural sociology, as occurred in the land-grant college system elsewhere in the US. The power of productionism downplayed the interests of the rural community and the University of California agriculture faculties were seen from an early stage as providing its scientific basis.

Despite this marginalization, there is a tradition of critical studies that dates back to the early twentieth century. For instance, Bailey (1915) ques-tioned the existing pathways of progress in industrial agriculture, arguing for a more welfarist approach to rural community development and for the preservation of the family farm as one bulwark of it. In his book *The Garden Lover*, he advocated the first principles of agro-ecology, later to be developed by Californian scholars such as Altieri (1988) and Gliessman (1990).

Bailey (1915) foresaw the need to reconnect the biological interdependence between people and nature and to resist monoculture and the decline of mixed farming, emphasizing, instead, the 'gardening' of horticulture. He also advocated a new set of connections between producers and consumers, with those increasing numbers who lacked the vital contact with their foods being 'standardised by the mere force of circumstances and imitation'. How-ever, Bailey's postulates, as Stoll (1998) reminds us, tended to fall on deaf ears, as they failed to inspire agri-business, public research, or policy at the time. Moreover, they were much vilified by other eminent agricultural scien-tists. Edwin Nourse (1924), for instance, an agricultural economist who spent a career in supporting the industrialization of Californian agriculture and became a leading and influential figure in US agricultural policy-making, made a stark distinction between the progressiveness of 'scientific farming' and the backwardness and sentimentality of Bailey's ideas.

Nevertheless, Bailey's ideas did survive these significant and periodic attacks. Writers such as Wendell Berry and Altieri and their students of agro-ecology sustained and built upon critiques of monoculture, reductions in genetic diversity, and single-crop farming methods. They saw these as hallmarks of the unsustainable industrialization of Californian agriculture, and as such, they considered them inherently self-defeating from an environmental and social point of view. As Stoll (1998) summarizes, single-crop farming stood as a critical focal point in this critique because it specifically related the natural advantages of the region to the institutions that growers and agri-business created to exploit them fully.

The critical 'tradition' was not just associated with a radical reinterpretation of the reconnection between agriculture and people. It also incorporated rural community development. Anthropologist Goldschmidt wrote a report (1947) for USDA which included two detailed case studies of Arvin, a community dominated by surrounding large farms, and Dinuba, a family farming community. He argued that 'the degree of urbanisation varies with the degree to which farm operations have become dominant'. Like his earlier colleagues, he dared to make a negative link between agri-industrialism and the rural community by focusing on the influence of the former on the latter.

Goldschmidt (1947) argued that in Arvin there were poorer conditions, a smaller middle class, lower family incomes, poorer public services, and less civic participation—what we might today term as a deficit in 'social capital'—and he directly related this to the large-scale and intensive agricultures. His findings were rubbished by the elite, both in California and in Washington. Even mentioning social stratification and community outcomes was seen as too destabilizing for the hegemonic agri-industrial paradigm. As Lobao and Meyer (2001) document, owners of large farms vented their anger by staged burnings of Goldschmidt's report and Steinbeck's *Grapes of Wrath* and launched attacks that eventually closed the USDA department to which Goldschmidt was associated. Thus, this controversy effectively extinguished any critical research on Californian agri-industrialism for over thirty years.

This lineage of struggle is significant today to contextualize current alternative agrarian movements in California. By dint of the political and economic strength of the agri-industrial complex described so far in this chapter, and indeed its very durability over time and space, these movements, more so than in other agri-food regions, are by definition and ideology oppositional. These were given a further and significant intellectual boost by the writings of Jim Hightower in the early 1970s, which, as Buttel (2003) points out, led to a renewed period of critical alternativeness for the next decade at least. Hightower (1973) called into question the role of publicly funded land-grant colleges, arguing that while originally designed to service farmers in their localities and communities, they had now become the handmaiden of agribusiness, helping seed firms and others to exploit the realm of agriculture.

As he explained (p. 57):

Land Grant college research for rural people and places is a sham . . . a look at the budgets and research reports makes clear that there is no intention of doing anything about the ravages of the agricultural revolution. The focus will continue to be on corporate efficiency and technological gadgetry, and the vast majority of rural Americans—independent family farmers, farm workers, small town businessmen and other rural residents—will be left to get along as best they can.

Hightower's radical critique, not just of the agri-business establishment but also of existing alternative voices, gave added impetus for alternative movements. However, the land-grant system was to be transformed by private sector interests in the 1980s and 1990s, when many potentially sustainable agricultural research schemes were terminated, as the rush to develop GM technologies, largely through closer engagements with the private seed companies, increased. According to Buttel, the 'Hightowerist' critique, based upon a representational alternative politics, has given way to a more fractured and diffused set of oppositions. These include a cluster of activism around GM development and the globalization of agricultural technologies and a cluster around sustainable agricultures and relocalization of agri-food. The latter, which we consider in more detail below, has increasingly focused on quasi-private efforts at developing community supported agriculture (CSA), green value-added labelling, alternative marketing strategies, and community food security. It has been in California where these 'new' agricultural movements were spawned in the 1990s, especially around the explosive growth of organics.

The New Frontier: The Explosive Growth of Organics as a Green Gold Rush?

By 1994 there were 4,050 certified organic farms in the US (Dunn, 1995). Organic sales grew by over $2.3 billion per annum, with an increase of 20 per cent per annum since 1989. The heartland of this growth was in California. With an emphasis mainly on salad mix, cotton, wine grapes, and a range of horticultural crops, the number of farms increased by 55 per cent between 1992 and 1995; by 1998, it was estimated that organic production constituted nearly 5 per cent of the total Californian agricultural economy. This explosive growth, as Guthman (2003) documents, was as much a result of the development of an urban-based 'yuppie' and 'nimby' counterculture as it was the latest example of the peculiar brand of production innovation which has characterized the state since the Gold Rush era.

What makes this development distinctive is the rise of a new productive and consumptive form which represents a highly 'contested space' between the competing conventional and alternative worlds of Californian food. As we shall see, this is a highly dynamic contested space played out partly through the development of standards and regulations regarding what actually constitutes organic food in the US. In this sense, it represents a new 'regulatory battlefield' around which an alternative agri-food paradigm becomes defined.

At first sight the rise of organics in California would seem to represent a major opportunity for embedding and spreading the principles of the agri-ecological and sustainable agricultural movements. In fact, it held the potential to re-establish Altieri's principles of 'farming in nature's image' (Altieri, 1988) and to link this directly with the explosive urban demand for organic goods, both in the home and in the growing number of chic organic restaurants. Guthman (2003), for instance, describes the rise in demand for the organic 'salad mix', where tropes of nature and health welded a connection between urban countercultures and the development of the first certified organic programme in the US (California Certified Organic Farmers, CCOF), which started from humble beginnings in Santa Cruz in 1973.

The Bay area, in particular, was an unusual haven for organic cuisine and food quality at a time (in the 1980s) when there was a national spiral of decline in food expectations and quality across the US, reinforced by the public effects of the Aldicarb and the Alar pesticide scares of 1986 and 1988, which saw a quadrupling of organic acres in two years (Schilling, 1995). Restaurant and domestic consumption of organic food was rising rapidly, as was the number of new entrants willing to convert to organic production because of the price premium. This represented a lucrative market niche, a new 'green gold rush', which potentially could have brought Californian agri-ecological principles into the mainstream.

A key feature of the rise of organics, however, has been an increasing diversity of production frameworks, types of producer, and working and marketing arrangements. In addition, it has been difficult for the traditional 'deep' organic proponents to maintain sufficiently robust production standards, given the growth both in production and consumption. In short, to use Storper's terminology, production standards have shifted from the ecological to the commercial conventions. With the insertion of corporate marketing through retail and restaurant chains and with strong and persistent 'regulatory attacks' from both federal and state agencies, there have been severe external and internal pressures on the process of commercialization, or what Guthman (2004) calls 'conventionalization' of Californian organics. It is worth specifying in more detail these external and internal contestations, as it is through them that we can shed conceptual light on the social and political struggles in a region that is dominated by a highly dynamic form of agri-industrialism.

External Regulatory Capture

Through its organic organization CCOF, California set up the first regulatory legislation to define organic products in 1990, with the California Organic Foods Act (COFA). This was similar to some of the non-organic, retailer-led private forms of regulation that have since come to dominate wider food chains (Marsden and Wrigley, 1995). In fact, while part of state policy, it is operated at arm's length and delivered by the industry. California organic regulation followed this model from the start, putting emphasis upon regulation of product rather than process, and giving plenty of opportunity for wide definitions of organic labelling in the marketplace. Indeed, the COFA establishes a legal baseline definition of organic growing practices that includes a list of allowable material that does not require third-party inspection or verification of practices and is enforced only in cases of confirmed violation.

As this regulation bedded down, there have been calls for more third-party certification so as to generate consumer confidence and to continue to differentiate organic goods 'in the marketplace'—increasingly a euphemism for supermarket shelves. This has allowed for marketized standards to be added to the baseline COFA regulation, thus placing much of organics in a highly competitive selling environment whereby value can be abstracted according to specific third-party standards set by private bodies. For instance, if a third-party certifier has been involved, growers can label their products as 'certified organic', rather than just 'organic'. Guthman (1998) documents how privately based third-party certifiers, such as Farm-Verified Organics and Quality Assurance International, have grown in the state and how they compete for market share and extend their regulatory conventions into the processing and retailing end of the chains.

There is a tendency for this hierarchy of standards to become not only marketized and competitive, but also scientifically and administratively driven. Guthman (1998) sums up the system:

all nine of the certification agencies doing business in California must take the COFA as a baseline, but each sets different standards for its member growers, follows different certification and enforcement procedures, and charges different certification and member rates. Most of the agencies refuse to release information about their standards, methods or members. Several of the agencies are for-profit organisations, and all survive on membership fees and assessments. This gives rise to potential conflicts of interest, in which protecting their members may be more beneficial to the agency than vigorously enforcing standards.

Despite their complexity, both state and private forms of regulation have allowed the development of input substitution on many organic farms (Allen and Kovach, 2000), which have departed from agri-ecology principles and fuelled the development of a vibrant organic inputs market.

The evolution of both state-wide and federal regulation has continually attempted to (1) broaden the definition of organic practices and products, and (2) restrict state-based regulation to a baseline upon which private certifiers can then compete and continue to differentiate the market. However, this is a highly contested political space. For instance, in 2000 the Federal National Organic Program Proposal to allow the 'big three' (i.e. GMOs, ionizing radiation, and sewage sludge) to fall within a further national regulation of organics was blocked by revisions as a result of much grassroot mobilization. Nevertheless, as we saw in Ch. 2, the emphasis is still very much on product regulation, rather than process, and on labelling and minimalist food safety rules (Goodman, 2000). This still sets organic regulation firmly within an 'industrial framework'; according to Goodman (p. 212), 'with the rules of the game definitively established, the formative industry will henceforth be exposed even more directly to the forces of capitalist competition and accumulation. An industry has been created from the fabric of a social movement, whose oppositional potential has been further attenuated and channeled towards the market visions of green consumerism.'

Deep and Shallow Organics and their Colonization by Agri-industrialism

By the 1990s, and armed with a particular combination of public and private systems of regulation, consumer demand for organics witnessed a 20 per cent annual growth in sales, and large-scale agri-business, composed of growers, processors, and retailers, made significant inroads (Buck, Fetz, and Guthman, 1997). Of the 1,533 growers in 1997, 76 held more than 1,000 acres in total crop production and produce on contract to the organic 'shippers'. Another group of organic farmers includes the ex-urban real-estate holders who have little interest in the ethics of 'deep' organic production. Guthman (2003) argues that today much of the organic industry is characterized by oligopsony, with a handful of very powerful buyers and hundreds of marginally committed growers who sell their produce to them. At the other end of the spectrum, there is a vibrant sub-sector of organic farms that market more or less independently and more directly to restaurants and supermarkets, as well as farmers' markets.

At one end of the production spectrum we then see the growth of organic agri-business, which is taking on the features of the conventional agri-food sector in California. Of significance here, as we saw also in the conventional dairy sector, were the increasing agricultural and real-estate land values, which forced producers to specialize and to intensify production on the most income-bearing crops, sending organic production down a similar path to that of conventional production. This has also diluted the agri-ecological basis of many organic practices. One of the best examples of this trend concerns the rising consumption of 'salad mix', and especially the use of 'baby salad greens' which are picked young and treated with soluble nitrogen (Chilean nitrate), which is known to kill soil micro-organisms and contribute to groundwater pollution. Salad mixes can be assembled like components from different farms and regions, with baby greens being able to be produced quickly and at a rate of several crops per year, using marginalized immigrant workers as happens in the industrial lettuce and strawberry industries (Friedland, Barton, and Thomas, 1981; Wells, 1998).

Large multinationals, such as Dole and the Californian-based Missionero and Earthbound, have moved into the shopping and selling of salad mix. By 2001, these firms had 7,000 acres in organic production (under the new partnership name of Natural Selection). While some of these firms started out as alternative initiatives, they have evolved into the newest form of agrarian capitalist development. Natural Selection, for instance, is the largest supplier of speciality lettuces and the largest grower of organic produce in the US.

These new organic variants of concentration, specialization, and intensification have also threatened the economic viability of those 'deeper' organic producers who have resisted such practices and continue to use complicated crop rotation practices, recycling all nutrients and relying on biological pest control. The more standardized, commercial, and label-oriented certification schemes tend to draw a veil over these deeper agri-ecological practices. As a result, in a neo-liberal market structure that prioritizes private systems of regulating, they create little or no market value beyond that created through 'short' and 'face-to face' producer–consumer reconnections, associated with such avenues as box schemes and farmers' markets.

In short, the prevailing regulatory conditions of the land markets and of the organic supply chains tend to marginalize 'deep organics', rendering them a niche system in a wider but shallower organic world of food. Deep organics thus become a form of 'militant particularism' (D. Harvey, 1996) through a particular brand of technocratic regulation that has managed to be considered as legitimate to consumers and profitable for agri-business (Goodman, 2000).

One consequence of the ways in which contested regulation of the Californian organic sector has evolved has been to create a severe problem of scaling-up *deep* organics into a transcending economic force. This, in turn, has created a new dualism in the organic sector between the *shallow/corporate* and the *deep/agri-ecological* groupings. This 'industry and niche' (Goodman, 2000) dynamic still holds contestation around the very definition and conventions of organic matter. A major axis of this contestation surrounds commercial versus ecological organic conventions. The spread of commercial conventions has led to a considerable 'colonizing neo-liberal discourse', based on technocratic criteria, which further fragments the wider and diverse sustainable agriculture movements and suggests that parameters other than organic may have a role. In the next section we turn to this wider but weaker niche of community and local agri-food movements.

The Refuge of Relocalization: Reinforcing Deep Organics as a Real Alternative?

In the context of this 'neo-liberal' mode of regulatory control of alternative agricultures in California (through one its main dimensions: organics), it is perhaps not surprising that a growing consensus has built up concerning the failure of the wider sustainable agricultural movements to articulate coherent strategies of engagement. This has continued to perplex many critical writers in the region, suggesting that even such movements have been too compliant and tolerant of the rationalist and technological sophistry of the recent organic experience. In short, agri-industrialism, supported by a new brand of public–private regulation, is seen as having made significant inroads into the sustainable agriculture movement; so much so that its critical trajectory has mutated away from its traditionally central issues, such as equity, food security, class and labour relations, sex, and race in agri-food networks (Allen and Sachs, 1991). Projects have lacked these socially transforming forces and have concentrated on the priorities of making agriculture sustainable (Allen and Sachs, 1991: 571).

However, such scepticism and pessimism in the Californian case needs to be critically examined. As Buttel (1997) argues, a significant strand of the environmental movement with regard to agriculture in the US has been the localizing tendency, or what he terms *alternative technologism*. Here, social movements are actively engaged in influencing public research institutions, as well as state bodies, to emphasize sustainable, low-input, or alternative agriculture. This can promote ecologically stable and socially more harmonious and decentralized agri-food networks. In the US, and in California especially, there have been many rapidly expanding community supported agriculture schemes (Kloppenburg, Hendrickson, and Stevenson, 1996), aimed at 'building community' through the production and circulation of deindustrialized and decommoditized foods, many of which are not necessarily certified as organic. How far has this type of movement gone in California? Can we see here a further alternative challenge to the dominant but mutating agri-industrial paradigm?

Alternative Food Initiatives and Community Supported Agriculture

Alternative food initiatives (AFIs), which include community supported agricultures (CSAs), originated in California in the 1960s. They were associated with civil rights movements, support for farm workers' rights, and the problems of food scarcity and poverty in urban areas. Alternative networks often tried to bridge these issues as well as viewing agriculture and food as a basis for empowerment of local communities. Their growth needs to be linked to the social character of California's agricultural labour force, which has relied heavily upon hired and temporary labour. Specifically, immigrant farm labour is built upon a low-wage economy. For instance, there were an estimated 700,000 farm workers in California seeking about 400,000 positions in 2000; this surplus in the labour supply created a downward pressure on wage rates and conditions (Inouye and Warner, 2001). A recent survey showed that 75 per cent of farm workers are paid an hourly average of $5.69. Three out of five farm-worker families live below the poverty level and three-quarters of workers earn less than $10,000 a year. Farm workers also come face-to face with the health risks of the agri-industrial model: between 1991 and 1996, there were 4,000 occupational poisoning cases reported (Reeves, 1999), which is probably an underestimate to the overall risks faced.

The earliest AFIs tried to contest these social and political conditions, and social justice was given priority over quality food. Organizations such as the Community Alliance with Family Farmers, Interfaith Hunger Coalition, and the Agrarian Action Project were established in the late 1970s and linked the need for an alternative agriculture to social justice and food scarcity. The universities and their surrounding neighbourhoods were a major territory for the development of these initiatives.

Allen et al. (2003*b*) have undertaken the most comprehensive survey of AFIs in California. They interviewed the leaders of thirty-seven Californian AFIs, including farmers' markets, urban growers' associations, organic and speciality products networks (see Table 5.1), and realized that they held a common interest in shortening food supply chains and empowering local communities. The researchers, however, noted also that there had been a significant shift away from an emphasis on social justice issues to questions of food access, urban community empowerment, and support for small farmers. In other words, while environmental issues continued to be important, the civil rights issues associated with agricultural labour were rarely mentioned.

In this sense, the AFIs had become more food-based and less socially oriented. Moreover, there were real internal tensions between different types of AFIs. For instance, those that addressed food access and food security issues tended to locate themselves within an alternative, rather than an oppositional, frame. Considering that it is difficult to be effectively oppositional unless one scales up from the local level and that there is a tension between 'militant particularism' and global ambition, they discovered that these organizations were often fractured, not only in terms of different substantive priorities, but also in the degree to which they were capable of dealing with the broader material and institutional issues in the organization of production and consumption.

Allen et al.'s report (2003*b*) is a telling contemporary story of the growth and ambition of AFIs and how these are differentially dealing with the dominant (agri-industrial) regional political economy within which they find themselves. They provide a cautionary tale of how such initiatives can become constructively marginalized in such a region, even when they experience growth in participation and are based upon an established history.

A particularly popular alternative movement set within this increasingly heterogeneous alternative framework concerns those initiatives which fall under the category of CSA. In these arrangements, farmers are committed to selling to the local communities usually through box schemes of organic fresh

Table 5.1. California agri-food initiatives

Organization	Location in California	Year founded	Programmatic focus
St Anthony's Foundation Farm	Rural northern	1956	Rehabilitation of low-income and homeless substance-dependent people
Davis Covered Market	Urban northern	1975	Farmers' market
Food First/Institute for Food and Development Policy	Urban northern	1975	A think-tank for issues of food and justice internationally, a membership organization
UC Davis Student Farm	Rural northern	1975	Agricultural education for university students
Berkeley Youth Alternative Market Gardening	Urban northern	1976	Youth services and rehabilitation
Southern California Interfaith Hunger Coalition (defunct)	Urban southern	1977	Inner-city hunger; set up farmers' markets in low-income urban areas
Community Alliance with Family Farmers	Statewide	1978	Statewide organization, precursor advocated for justice for farm-workers but current organization is more focused on agricultural environmental issues, economic opportunities for family farmers
Common Ground Garden Program	Urban southern	1978	Urban agricultural education and access to urban gardens for low-income communities in Los Angeles
Berkeley Farmers' Market	Urban northern	1981	Farmers' market
'Heart of the City' Farmers' Market	Urban northern	1981	Farmers' market in the inner city

Table 5.1. Continued

Organization	Location in California	Year founded	Programmatic focus
San Francisco League of Urban Gardeners	Urban northern	1983	Urban economic development, community organizing, and empowerment
Richmond Farmers' Market	Urban northern	1984	Farmers' market in a low-income area
Select Sonoma	Rural northern	1988	Regional label
Santa Cruz/ Watsonville Farmers' Market	Urban central	1989	Farmers' markets
Homeless Garden Project	Urban central	1990	Rehabilitation and support of homeless people
Arcata Educational Farm	Rural northern	1992	Agricultural education, CSA
California Food Policy Advocates	Statewide	1992	Food policy, food access
Food from the Hood	Urban southern	1992	Urban agriculture, microenterprise for scholarships for low-income youth
Humboldt Harvest	Rural northern	1992	Regional label
San Francisco Jails Project	Urban northern	1992	Rehabilitation of people in jail
Occidental Center for Food and Justice	Urban southern	1992	Policy and programme development for inner-city food needs
Berkeley Opportunities for Self-Sufficiency	Urban northern	1993	Food security, community economic development, community gardens
Center for Urban Education about Sustainable Agriculture	Urban northern	1993	Urban agricultural education, farmers' market
Center for Urban Agriculture at Fairview Gardens	Urban southern	1994	Demonstration organic farm, urban agricultural education
Long Beach Organic	Urban southern	1994	Community gardens for urban poor
Occidental Arts and Ecology Center	Rural northern	1994	Agricultural education, lifestyle change, international community

Organization	Location in California	Year founded	Programmatic focus
Center for Agroecology and Sustainable Food Systems CSA	Urban central	1995	CSA associated with Apprenticeship in Ecological Horticulture, a practical education programme
Marin Food and Agriculture Project	Urban northern	1996	Regional food security policy
Park Village	Urban northern	1996	Community gardens for residents of public housing
Berkeley Community Gardening Collaborative	Urban northern	1997	Community gardens, community food policy
Berkeley Food Systems Project	Urban northern	1997	Agricultural education, alternative markets to schools, hunger issues
Escondido Community Health Center	Urban southern	1998	Community gardens for Latino residents
Amo Organics	Rural central	1999	Latino farmer marketing cooperative, CSA
Yolo/Davis/Winters Farm to School Project	Urban northern	2000	Farm to school programme
Community Food Security Coalition Farm to School Project	Urban southern	2001	Food and agricultural education, farm to school programme

Source: Allen et al. (2003*a*).

produce. Residents often become shareholders, in the sense that they pay a stake earlier so as to spread the risks of investment and planting. The idea began to develop in the US in the 1980s and there are now nearly 1,000 CSA farms in the country. This is seen by the producers and the communities as a way of keeping small farms alive and reconnecting the production and consumption spheres.

Some detailed examples come from the counties of Alameda and Stanislaus, which have been studied in "FoodShed" projects based in the University of California Sustainable Agriculture Research and Education Program (Anderson, Feenstra, and King, 2002; Cozad et al., 2002). *Alameda*, located in the heart of the Bay area, had developed nineteen farmers' markets by 1999 as well as twelve subscription 'food baskets'/networks (CSAs), four organic distribution services, and four roadside stands, giving local consumers real alternatives to the main grocery outlets. In this growing 'Bay area culture', which is seen to nurture direct marketing, the consciousness about problems of the agri-food system are rising (Belasco, 1990) and this, as in parts of Europe, is creating new connections between the urban and the rural. There are significant threats to this development, however.

Rapid suburbanization is taking up agricultural land and putting up its price on the one hand, but it is also providing potential growing markets for CSAs on the other. Many farms are filling up to 200 food baskets weekly and some farmers are orienting over 50 per cent of their production through this route. A major challenge is to keep the supply of foods local, given the growth in demand and the disappearance of small local farms due to real estate development. There is a danger that such localization initiatives become stretched, partly because the demand for the foods increases and it can only be met by non-local suppliers.

Stanislaus County is in the heart of the Central Valley and it is one of the state's most fertile regions. Between 1945 and 1997 the number of farms under 50 acres fell by 45 per cent while the number of those over 100 acres remained constant. As with Alameda, the county is experiencing rapid population growth with suburbanization. The population increased by 116 per cent between 1970 and 1997, with an average of 600 acres of farmland being converted to development (mainly residential) every year between 1984 and 1998. Despite largely middle-class suburbanization, 18 per cent of all residents and 27 per cent of children live below the poverty line.

CSA arrangements and farmers' markets are less developed, with only two farmers' markets in the county. Fewer people are willing to pay a premium for local produce and farmers seem also less interested in innovating with new, more direct marketing arrangements. National restaurant and retail stores are more tied to the uniform national distribution channels and this creates a lock-in effect for farmers. Where direct marketing is occurring, it is creating greater incomes for farmers and there is a need to develop these initiatives from a lower base than in Alameda.

Perez (2004) surveyed 274 CSA members. They were shown to have positive advantages with regard to producing high-quality produce, assisting consumers to develop healthier eating habits, and reducing chemical and fuel usage on farms. There are several challenges that can affect the long-term viability of such initiatives, not least the fact that the actual local geographical context becomes a key issue in the evolution and potential take-off and clustering of these activities. The level of convenience and choice also becomes an issue, even if consumers are prepared to pay more.

The Arrested Development of Organic and Alternative
Food Initiatives in California

Even if many local and organic initiatives seem to deradicalize as they mature, it is clear that they can begin to represent real alternative innovations in producer–consumer relations. We see in Alameda the development of a clustering that is creating new webs of producer–consumer relationships outside, and significantly autonomous from, the prevailing conventional system. This places emphasis on their key actors to manage the process of relocalization in ways that sustain these webs while coping with the external changes in demographics and the price discounting tendencies operating in the conventional retail-led system.

While there may be, for the purists, considerable variation in the degree to which such local initiatives are achieving the goals of a more ecological paradigm, it is clear that new suburban–rural alliances and reconnections are developing and changing the agri-food landscape in some places. This represents a sort of alternative and relocalized archipelago in a sea of mutating agri-industrialism. As Guthman (2003: 50) says, maybe there is some salience in the organic boosterism displayed in this telling statement: 'salad mix [consumption] has done more to reduce pesticide use in California than all the organizing around pesticide reform'.

The evidence from the Californian alternative world of food suggests that, while the agri-industrial paradigm is quite successful at overcoming the new obstacles it faces with regard to the organic consumption explosion, those initiatives that prioritize a relocalization strategy represent a far tougher challenge. The agri-industrial paradigm, aided by an ideology of neo-liberal markets in land and food goods and private systems of quality regulation, may be able to render organics as little more than a discounted form of pesticide reduction and revalorization. However, it has much more difficulty in challenging the reflexive localization tendencies of a myriad of alternative farm-to-consumer short circuits. These can display much more autonomy, and are less regulatable in a traditional sense. It remains to be seen, therefore, what might happen if they were to scale up and attempt to keep their local provenance. Again, the operation of the land market seems to be a factor in making this difficult to achieve.

Nevertheless, precisely because local AFIs tend to play by their own rules, rather than by those managed by the state and other private-sector interests, there is justification for being more optimistic about their trajectory as self-sustaining. Moreover, as the low-wage economy and the poorer quality of foods extends from the farm to the corporate retailer in the conventional sector (as demonstrated by the public concerns about Wal-Martarization in the US), further public and consumer concern may develop. This could further drive a shift to 'buying local'.

Conclusions: The Battleground between Two Rival Worlds of Food

In the introduction to the chapter, we raised questions concerning the relationships between the two rival worlds of food in California. Our analysis here has shown that the dominant agri-industrial world is far from static and that it faces a major challenge with regard to the onset of 'third nature' concerns. At the same time, the fragmented and diverse alternative model has gained public and market approval, and a major battleground is located around the regulation and definition of organics. One dominant thesis here concerns the colonization and conventionalization of organics as well as some of the sustainable agriculture movements. 'Sustainable agriculture' is a term used throughout the Californian governance framework and is a major plank of the research and development work in universities.

On the other hand, different forms of relocalization of food are also taking place at the interstices of the conventional system. It is unclear how these rival worlds will play out. At the moment, there seems to be a sort of coexistence, which reflects the contradictions of agri-industrialism on the one hand and the more 'reflexive consumer' on the other. The former continues to strive for accumulation and public legitimacy, while the latter, tied now to a highly mobile consumer culture, wrestles with the choices of cheap and premium food purchasing.

The continued dominance of the highly dynamic and mutating agri-industrial model in California has also spawned an oppositional and alternative food culture as the population has expanded and suburbanization has continued apace. The two rival worlds (depicted conceptually as a battleground in Ch. 3) are in many ways trying to colonize each other as well as protect their distinctive characteristics. The design and implementation of 'quality' regulation becomes a key but potentially movable outcome of this battleground, whether it is associated, as we have seen here, with strawberry pesticide policy, organics certification, or intensive dairy production. In short, *relocalization* challenges the traditional forces of *agri-industrialism*, while the latter makes the former all that much harder to sustain.

6

The Commodity World in Wales

As the first industrial nation, the UK was one of the earliest countries to experience the industrialization of agriculture, a process that led to an unprecedented increase in productivity, with more and more food produced by fewer and fewer people. Early exposure to intensive food production clearly left an abiding cultural legacy; to this day, one of the proudest boasts of the British food industry is that it renders cheap food to the consuming public at ever lower prices. This production ethos was both cause and consequence of a mainstream consumption culture which sets a high premium on price and treats food more as fuel than as pleasure. In his thousand-year history of British food, Spencer (2002) caught this aesthetic perfectly when he suggested that the British 'were unexcited by the food they ate, but they knew that they had to get on and eat the wretched stuff'.

In its attachment to cheap, processed food, the UK is far closer to the US, the quintessential fast-food nation, than to Italy, France, or Spain, countries where there continues to be a strong cultural appetite for fresh, local, and seasonal food. Although Britain's cheap-food culture has complex and manifold causes, its origins lie in the early period of industrialization, especially in the system of colonial preferences from the Commonwealth countries, which created a low-cost template for locally produced food. In other words, the global–local interplay that did so much to shape economy and society in Britain also influenced the economics of food production and the culture of food consumption.

To a greater extent than in other European countries, the supermarkets have become the key players in shaping food consumption patterns in the UK. As in California, retailer power is now the key to understanding the enormous asymmetries of power that punctuate the British agri-food chain from farm to fork. One reason why supermarkets seem to wield so much more power in the UK than their analogues in other countries is that there is less countervailing power at the production end of the UK food chain. In contrast to other European countries, where farmers seem to have forged durable and robust cooperative structures, British farmers have been unable or unwilling to sacrifice their individual identities for more collective power as producers. This 'possessive individualism' is beginning to exact a terrible price. Indeed, as this chapter will show, the failure to aggregate their power as producers and to act in concert in the food chain is probably the single biggest political failure of British farming in the twentieth century.

The commercial costs of this political failure were obscured and mitigated until recently by the Common Agricultural Policy (CAP), a system that subsidized farmers as individual producers and, therefore, helped to underwrite the self-indulgent and short-sighted culture of possessive individualism. However, as we saw in Ch. 2, the recent reform of the CAP signals the end of production-related subsidies, with the result that UK farmers find themselves more exposed to the commercial pressures of their customers, especially the supermarkets, than their EU counterparts, many of whom can draw on cooperative farming structures to help them withstand these new pressures.

The reform of the CAP—a belated and convoluted process that reregulates, rather than deregulates, the agricultural support system in Europe—poses two major questions for UK farmers. First, can they transcend their traditional culture of possessive individualism in favour of more cooperative governance structures? Second, can the less favoured areas of the country, which could never compete on price, make the transition from the world of commodity producer to the world of quality producer? In this chapter we shall address these twin challenging issues in the context of Wales to illustrate the wider dilemma of small nations and regions in the EU that are not just trying to upgrade their agricultural sectors in value terms, but are endeavouring to do so in the context of a political commitment to sustainable development. Specifically, we will base our analysis on the following themes:

• section two sets the scene by outlining the nature and implications of the *commodity ghetto* in Wales, highlighting the economic and political reasons as to why this region came to be 'locked in' to a narrow, path-dependent trajectory of low-value commodity production;

• section three analyses the two main types of agri-food chain—dairy and red meat—exposing the common threads running through these chains, such as the weak bargaining position of fragmented producers and the lack of producer-controlled processing capacity;

• section four examines a new agri-food strategy which is emerging as part of a wider political commitment to sustainable development in Wales. As we will explain, this agri-food strategy aims at a cultural revolution by fostering more cooperation among producers, more high-value branded products and more discerning local markets to stimulate the demand for new products;

• finally, section five assesses the prospects for a successful transition from one 'world of food' to another in the context of a more devolved governance system.

The Commodity Ghetto in Wales

The UK's agri-food system may be highly industrialized but it is far from being uniform across the country. Among other things, there is a prominent spatial division between the western and eastern regions, with the former geared to livestock and the latter to arable production. No less striking is the division between the uplands and the lowlands, with much of the former classified under the EU designation of Less Favoured Area (LFA). Within these broad patterns of spatial differentiation, there are other, more recent distinctions, such as the uneven development of organic agriculture and the markedly different attitudes to the use of GM technology. However, nothing better illustrates the variegated nature of farming in the UK than the uneven geography of LFA status, which concerns 10 per cent of England's agricultural land, compared with 70 per cent in Northern Ireland, 77 per cent in Wales, and 84 per cent in Scotland (Ward and Lowe, 2002).

An unfavourable combination of soil, terrain, and climate makes Wales a very difficult area to farm. For this reason, much of the country falls under the LFA designation. These natural conditions also help to explain the narrow sectoral patterns of specialization. With some 10 per cent of the UK dairy herd, 13 per cent of the UK beef herd, and over 25 per cent of all UK sheep, Wales is virtually synonymous with the livestock sector, to the point where livestock products accounted for 85 per cent of total Welsh agricultural output in 2002 (WDA, 2004).

Many of these mainstream products—milk, beef, and lamb especially— have remained trapped in the low value-added commodity category because, until very recently, there was little or no concerted attempt to engage in serious product branding exercises or producer-controlled processing activities. This, in short, is what we mean by the *commodity ghetto* in Wales. Breaking out of this commodity ghetto is now perceived, by politicians and the industry alike, as the key challenge for Wales in the agri-food sector. The new strategy, which we will examine later, is brutally clear about the nature and the significance of this issue (WAG, 2001):

the fundamental choice facing the industry in Wales is whether to continue to try to compete in the markets for basic agriculture and food commodities, where competition is on price, or whether to move as far as possible along the spectrum towards competing less on price and more on quality. The latter is the only realistic option if the objective is to try to slow the decline in agricultural employment and help as many family farms as possible to survive. This means developing high quality, value-added, branded products, which are aimed, where possible, at more specialised markets and niche markets.

Before examining the key agri-food chains in more detail, it is worth asking: why has Wales been locked into this narrow, path-dependent process of basic commodity production for so long? Why, in other words, was so little done in the past to build brands, to form production cooperatives, or to create producer-controlled processing capacity? To begin to answer this question, we need to examine the basic unit of production in Welsh agriculture: the family farm.

In contrast to the large 'barley barons' of eastern England, where farms have evolved into large-scale agri-businesses, the small family farm remains the dominant social unit in Welsh agriculture. This distinctive farm structure assumes a relevant cultural significance; in fact, considering that more than half of all farmers speak Welsh, the family farm in Wales helps to define the character of rural society and its sense of identity—an identity that may be threatened by the seemingly ineluctable growth of larger farming units. The family farm might enjoy an iconic status in terms of Welsh cultural heritage, but in purely economic terms it has its limits as a business vehicle for capital accumulation. This factor may help to explain the short-term, penny-pinching attitude to investment beyond the farm, in production cooperatives and joint marketing initiatives, for example.[1]

While the small family farm, and its limited economic resources, can help to explain the stubborn longevity of the commodity ghetto, it cannot carry the full burden of explanation. Indeed, other factors conspired to the same end. These include:

- the culture of possessive individualism, which worked against producer collaboration;
- the political failure to forge a common voice for Welsh farmers as a result of internecine conflict between two rival trade unions;
- the growing power of retailers and processors, which made it even more difficult for producers to engage and become involved in vertical integration;
- mainstream consumption patterns in Wales and the UK, which generally offered limited opportunities for developing high-quality food products;
- the continued flow of CAP subsidies, which rendered the unfavourable status quo viable.

Far from being confined to Wales, the culture of possessive individualism is a British phenomenon, partly induced by the fact that overall farm structures in the UK were larger than in continental Europe, enabling farmers to remain independent of one another. This ethos was also fuelled by the National Farmers' Union (NFU), which tended to be unduly influenced by the ideological interests of the larger farmers in England, as we will discuss later. Possessive individualism may have been more muted in Wales, where farms were so much smaller than in England, but it took root nevertheless.

---

[1] The small scale of the Welsh family farm is not a sufficient explanation for the historical aversion to cooperation and cooperatives. In fact, many continental European farms are even smaller, but this has not prevented them from forming cooperatives—indeed, it has been interpreted as an incentive to cooperate with others.

As one influential Welsh farmer put it (R. Roberts, 2001): 'I cannot pretend that it has been easy, within the industry, to engender enthusiasm for co-operation, many farmers feel that there are perhaps too many stories of failure.'

The story generally believed to symbolize the failure of the cooperative principle in Wales concerns the fate of the co-op Welsh Quality Lambs (WQL). Created in 1970 as a collective vehicle to improve the marketing of Welsh lamb, WQL did exactly what farmers are enjoined to do today, namely, to engage in vertical integration. In 1981, WQL purchased a controlling stake in a meat-processing business, which included a high-grade abattoir facility. This move eventually brought WQL to its knees; as a post-mortem explained (Co-operative Development Board, 1987), 'the major decision which led to the eventual demise of WQL was to invest in an under-capitalised business without proper evaluation of its commercial viability'.

A shortage of capital, combined with an alarming dearth of business acumen, did more than simply destroy a thriving marketing co-op. Far more damaging in the long term was the fact that Welsh farmers drew the wrong lesson from this experience, as they came to believe that WQL furnished incontrovertible evidence that cooperative businesses were doomed to failure. Today, it is generally accepted that the real lesson was that co-ops need professional business skills, especially marketing skills, if they are to avoid the tragic fate of WQL.

In addition to the effect of stories of failure, divisive trade union politics also hampered cooperation among farmers. The origins of this situation lay in radically different political assessments of how best to cater for the distinctive interests of the small family farm in Wales. For years, many Welsh farmers felt that they were underrepresented in the key NFU committees that negotiated price reviews and marketing terms, as the NFU leaders were inordinately often drawn from the ranks of the large English farmers. The issue which seemed to symbolize the big-farmer bias of the NFU was the proposed flat rate increase in union subscriptions, which asked the 80-acre Welsh farmer to pay the same amount as the 2,000-acre English farmer. These tensions eventually triggered a revolt in Carmarthenshire on 3 December 1955 which resulted in the formation of the Farmers' Union of Wales (FUW). Although the FUW had to wait until 1978 before the government recognized it as the equal of the NFU as an interlocutor, for decades Welsh farmers remained ideologically split into two rival unions, with the FUW playing the role of a vociferous champion of the small farmer.

The most important reason why the commodity ghetto went unchallenged for so long was that there was little or no incentive, from either the market or the state, to challenge the status quo, since it provided a modest but tolerable existence for the vast majority of farmers in Wales. The commodity production ethos on the supply side was perennially reinforced on the demand side.

In fact, as we explained earlier, the main food consumption trends in the UK market were stubbornly oriented to the commodity end of the food chain. This was especially the case in Wales, where, in stark contrast to Tuscany, for example, a noxious combination of low family income and poor diet furnished little or no demand for locally produced or regionally certified produce.

If the market afforded few incentives to challenge the commodity ghetto, the state was even more of a conservative force on account of the CAP subsidies that cosseted the industry, rendering it impervious to the need for change and innovation. It is no coincidence that a radical strategy for breaking out of the commodity ghetto in Wales appears only when the constellation of factors which had sustained the status quo begins to unravel. In particular, the crisis in the commodity markets, the glaring inequalities of power in the food chain, the reform of the CAP subsidy system and the advent of a directly elected Welsh Assembly together provided a new set of threats and opportunities.

Supply-Chain Asymmetries: The Plight of the Primary Producer

CAP reformers invariably claim that the decoupling of subsidies from production, the centrepiece of the reform, will allow the market, and not the state, to determine the level and quality of what is produced. However, as we shall see, for many primary producers in the UK the 'market' means not the anonymous and well-balanced forces of a neoclassical mechanism in equilibrium, but a profoundly imbalanced supply chain in which a small number of retailers exercise ever more control over what is produced, where, and by whom. Although there are many routes to market—farmers' markets for example, which are growing in number, albeit from a very low base—the multiple retailers are the single most important route, considering that the top five account for over two-thirds of all retail grocery sales in the UK.

The growing power and influence of the multiple retailers presents a stunning contrast to the declining power and influence of the primary producers. To illustrate the problem of supply-chain asymmetries, and to highlight the plight of the primary producer in particular, this section focuses on the three main sectors of Welsh agriculture: dairy, lamb, and beef.

The UK is the seventh largest milk producer in the world and the third largest in Europe, after Germany and France. The raw milk produced by UK dairy farmers is processed into a number of different products, the most important of which are liquid milk and cheese, which accounted respectively for nearly 50 and 25 per cent of total UK dairy production in 2002.[2] While a

---

[2] Unless otherwise stated, our analysis of the dairy sector relies on the evidence submitted to the House of Commons (2004) inquiry into milk pricing in the UK.

significant proportion of the commodity products—such as cheese, butter, and milk powder—is exported, liquid milk itself is sold almost exclusively in the domestic market. It is worth underlining that, as a natural product, milk is unique. In fact, it has to be processed within 24 hours and its supply cannot be stopped in the short term. These natural attributes of the product carry important implications for the organizational structure of the dairy industry because milk production needs to be closely coupled with milk processing—a link that has generally promoted a high level of vertical integration in this industry around the world.

Vertical integration may be the norm elsewhere in Europe, where producer-owned cooperatives have moved into processing, but a distinctive feature of the UK market is that a mere 10 per cent of raw milk is processed in this way, with the remaining 90 per cent processed by privately owned dairy companies. Among other things, this means that the profits from processing are passed back through dividends to dairy shareholders, rather than through milk prices to producers, as happens in many other European countries.

The structure of the UK dairy sector, where producer-owned cooperatives are predominantly brokers rather than vertically integrated processors, represents an organizational weakness that imposes a very heavy burden on producers. For example, the low level of vertical integration is considered one of the main causes of farm gate prices being the lowest in the EU. A thorough analysis of the UK dairy chain (KPMG, 2003) identifies four fundamental problems with the market, namely:

- supermarket power;
- the bias to commodity products;
- sectoral inefficiencies;
- regulatory environment.

Since these problems go a long way to explain the plight of the primary producer, let us briefly elaborate on each of them.

Supermarket Power

According to the management consultancy KPMG, the growing share of dairy products sold through supermarkets enhances the market power of the multiple retailers, with the result that a smaller proportion of the retail price is passed back to dairy farmers. Such enormous retail power generally increases the speed at which farm gate prices follow the market down and, conversely, it slows the speed at which they follow the market up. This trend has recently led the House of Commons to conclude that the balance of power between supermarkets and primary producers is heavily weighted in favour of the former and that this uneven distribution of power helps to explain why the dairy market is 'slow to react to upward pressure on retail prices' and why prices increase only 'following direct action by farmers' (House of Commons, 2004). Supermarket power also helps to explain why retailers' margins on liquid milk and dairy products have increased significantly over the past decade (with the gross margin on cheddar cheese, for example, rising to as much as 60 per cent), while farm gate prices and margins have continued to fall (Milk Development Council, 2004). These radically different trends in prices and margins between primary producers and supermarkets have fuelled calls for a major review of the code of practice which is supposed to regulate the way supermarkets deal with their suppliers—a code that currently lacks any statutory power (Marsden, 2004*b*).

Commodity Bias

The second problem identified by KPMG is an overdependence of the dairy market on commodity-type products, along with a lack of innovation and a poor marketing record. This means that a litre of UK milk generates a significantly lower return at retail and wholesale level than a litre of milk in other EU countries. The commodity bias also helps to explain why the UK trade deficit in dairy products increased by 60 per cent between 1998 and 2003. In fact, cheese imports, for example, consist of high-value specialist products, while cheese exports consist of low-value commodity products. Significantly, the average value of a tonne of imported cheese is £1,400, compared with £950 per tonne for exported cheese (Milk Development Council, 2004).

Sectoral Inefficiencies

One of the main problems identified by the KPMG report is that the industry, at both processing and production levels, is not very efficient. This problem affects primarily the processing sectors but it also exists at the farm level, where there is a large productivity gap between the best and the worst dairy farmers.

Regulatory Environment

The final problem concerns the regulatory environment in the UK, where one of the chief issues is the uneven application of competition law in the UK and EU authorities. The report anticipates a major restructuring in the dairy industry and it argues that competition authorities in the UK ought to take a sympathetic view of this process and assess it in a broad European context, rather than in a narrow UK context.

This last point assumes enormous significance in the light of the regulatory (mis)management of the UK dairy industry. Following the deregulation of the UK dairy industry in 1994, only 50 per cent of milk producers stayed with the cooperative structure; the other 50 per cent took the option to supply their milk directly to the processors, with the result that dairy farmers lost a united voice. The plight of the primary producer got even worse when Milk Marque, the largest dairy co-op, felt obliged to break itself up after being accused by the Competition Commission of abusing its 'monopoly position'—even though its members accounted for only 36 per cent of total UK milk production.

In retrospect, the break-up of Milk Marque is highly anomalous, given the general trend in the UK towards growing concentration among processors and retailers. This is also true when comparing the UK situation with producer trends elsewhere in the world: in Denmark and Sweden, for example, Arla Co-op collects, processes, and markets close to 90 per cent of the milk, while in New Zealand Fontera collects, processes, and markets as much as 98 per cent of the milk. Clearly, dairy producers in these countries have benefited from a more benign regulatory regime, characterized by less strict competition laws compared with their UK counterparts.

The plight of the UK dairy farmer could deteriorate even further if the worst scenarios emerge from the 'milk war' precipitated by the big retailers in the summer of 2004. The supermarkets, which account for some 62 per cent of total milk sales in the UK, began to restructure their milk supply chains, with major consequences for the three biggest dairies and the primary producers who supplied them. The early skirmishes in the new 'milk war' took the following form:

- 25 May: Asda (owned by Wal-Mart) picks Arla as its sole milk supplier, squeezing out the other two dairies, Robert Wiseman and Dairy Crest;
- 25 August: Sainsbury's squeezes out Arla in favour of Dairy Crest and Robert Wiseman as its long-term suppliers;
- 27 August: Tesco cuts its suppliers to two, terminating Dairy Crest's £60 million p.a. contract.

Although the dairies were the first to feel the pain of the new milk war, primary producers were rightly anxious too. In fact, price cuts tend to be passed down to the farmer, which is precisely what happened when Arla, the biggest dairy cooperative in the EU, reduced its farm gate price by 0.4p a litre.

Supply-chain pressures on farm gate prices could be exacerbated by regulatory pressures in the shape of CAP reform. Dairy premiums have been decoupled from production and incorporated into the new Single Farm Payment (SFP), which began in January 2005. In the short term, the farm gate price is expected to fall, perhaps to as low as 15p a litre.

The Milk Development Council predicts that a 25–30 per cent reduction in milk supply would end the production of commodities, forcing the UK to import dairy products (other than liquid milk) from other countries. Processors would then have to pay higher prices to foreign producers to secure sufficient supplies of raw milk for liquid milk production. In the worst case scenario, the combination of supply-chain pressures and CAP reform could induce a major shake-out of UK dairy farmers: from a current base of some 25,000 to just 15,000 in 2015. Such a drop could cause a major shortage of raw milk, though this would also ensure higher farm gate prices for the dairy farmers who survive the restructuring process (House of Commons, 2004).

Supply-chain asymmetries are also very apparent in the red meat sector. In 1996, the UK government conceded that BSE could be transmitted to humans, and this announcement quickly decimated the beef trade. In 2001, before the beef market had fully recovered, the UK was hit by foot-and-mouth disease, an epidemic which led to the mass slaughter of more than 6 million animals before disease-free status was eventually regained on 21 January 2002.

The livestock farming crisis in the late 1990s was especially acute in Wales, where the Welsh Affairs Select Committee of the House of Commons launched an inquiry to examine the growing disparity between farm gate and retail prices, following charges that the big retailers were 'profiting out of a crisis' (Marsden, 2004). The Select Committee concluded that the 'producer had borne the brunt of the reduction in returns' and called for an independent study of the retail pricing of meat products (House of Commons, 1997). Such a study was eventually undertaken by the Competition Commission, which concluded that 'excessive prices are not being charged, nor excessive profits earned'. However, the supermarkets were not completely exonerated: of fifty-two alleged unfair trading practices, the big retailers were found guilty of practising most of them, including asking their suppliers for non-cost-related payments and discounts, imposing charges, changing contractual arrangements without adequate notice, and unreasonably transferring risks to their suppliers (Competition Commission, 2000).

Primary producers had to wait another year before the Commission's main recommendation—a legally binding code of practice to govern the relationships between retailers and suppliers—was published on 31 October 2001. Two aspects of this code have attracted particular criticism from the farming unions. First, the code only applies to the top four retailers with market share of 8 per cent and above, which means that the thousands of producers who supply the other supermarkets are not afforded any protection at all. Second, the modest protection offered by the code is more apparent than real because, however unfair the practice, suppliers are very often reluctant to complain for fear of being delisted from the retailer's supply chain. Taken together, these shortcomings devalued the practical significance of the code in the eyes of the primary producer.

Despite these supply-chain problems, the relationship between producers and retailers in the red meat sector has gradually improved. In contrast to the dairy sector, where there is a chronic surplus of liquid milk on the market, demand for high-quality red meat outstripped supply in the wake of the foot-and-mouth epidemic, with the result that farmers have enjoyed better prices in meat than in milk. In this relatively benign commercial environment, one of the most significant trends is the growth of retailer-led producer clubs in which supermarkets and their dedicated suppliers enter a long-term partnership to ensure a consistent supply of high-quality meat for the mutual benefit of each side of the supply chain—a rare positive sum game in the UK context.

From the retailer's standpoint, these producer clubs are a good example of enlightened self-interest. Retailers need a guaranteed supply of quality meat from farm-assured producers who can satisfy the consumer's growing desire for a fully traceable product. By appreciating the need for stable sources of supply, retailers like Waitrose, Tesco, and Sainsbury's are not seeking 'ever lower prices' from their suppliers as they know that this would threaten their economic security.

From the farmer's standpoint, the producer club also makes commercial sense. With the growing commercial pressures of CAP reform, livestock farmers are reluctant to expose themselves to the vagaries of the market after bearing all the costs of raising their stock. Younger farmers tend to be the more receptive to the assured prices of the clubs, partly because they are more business conscious and partly because they are more financially in-debted than their elders. Older farmers, in contrast, have shown a remarkable degree of loyalty to the traditional auction markets, which they see in part as social occasions, rather than purely economic transactions.

In volume terms, the red meat supply chain in Wales is dominated by retailer-led producer clubs, the largest of which is controlled by Tesco, the UK's biggest supermarket. Borrowing from the 'lean' supply chains in the auto industry, where pioneers such as Toyota use first-tier suppliers to manage lower-tier suppliers on their behalf, Tesco uses St Merryn Meats (SMM) to manage its red meat supply chain throughout the UK. Since arriving in Wales in the mid-1990s, SMM has grown to such an extent that it is now responsible for processing half of all lamb and beef produced. SMM's high-technology plant at Merthyr, in the South Wales valleys, is a state-of-the-art facility, incorporating the latest capital intensive features. One of these features is the use of chill (not frozen) hanging techniques, with the carcass hung at $+/-0.5°C$ for an unspecified period of time (as this is considered a trade secret) to improve the flavour. Another is the development of modified atmosphere packaging (MAP) techniques that extend the shelf-life of the meat from 2 to 7 days and help to maintain its appealing red colour.

These high-technology features are demanded by Tesco, which sources two-thirds of its lamb and three-quarters of its beef from SMM in what appears to be an extremely tight supply chain. SMM is obliged to use abattoirs that have been approved by Tesco and all the members of its producer club must be farm assured. Under these tightly controlled supply chain regulations, SMM will contract farmers to sell livestock at specified times of the year, requiring a specific quantity and a certain quality level. Since some livestock farmers in Wales, and in the UK generally, have had problems in meeting retailers' carcass specifications, SMM provides incentives (and penalties) as part of its 'tutoring' programme to help farmers to identify the required carcass specification, which consists of a combination of size and fat content. SMM also convenes 'open days' for its producer club members, so that farmers can see the variability of quality on the hook and learn how to meet carcass conformation standards.

Traditionally wary of such arrangements, livestock farmers are coming to accept these retailer-led producer clubs as a more viable and predictable alternative to the declining returns that they were receiving on the open market. The advent of these tightly controlled supply chains signals a significant shift in the traditional system of meat procurement in Wales. In fact, in the past the industrial food chain was associated with anonymous, placeless products, whereas the most important feature of red meat marketing today is traceability, particularly the need to establish a link between region of origin and consumer perceptions of quality. Even SMM, the acme of productivism in the UK, feels obliged to brand Welsh produce as such.

Aside from top-down, retailer-led initiatives such as the Tesco example, there have been a number of bottom-up, farmer-inspired producer clubs. One of the most successful examples is Lleyn Beef, founded in 1997 by a group of forty beef farmers in the Lleyn peninsula, in North Wales. The main aim of this producer group is to add value to their high-quality beef by identifying specialist markets; to this end, they sell to local butchers, hotels, and restaurants, using a strict traceability protocol and membership of Farm Assured Welsh Livestock as a condition for becoming a member of the group. Today the Lleyn Beef producer group has a membership of some 250 beef farmers and its products are marketed nationally through retailers and catering butchers. Though still very successful, Lleyn Beef suffered a terrible blow when Cwmni Cig Arfon, a locally based farmer-owned meat company which bought 3,000 cattle a year from the group, went bankrupt in mysterious circumstances. Evoking memories of the Welsh Quality Lamb debacle twenty years earlier, the failure of Cwmni Cig Arfon, which collapsed owing nearly £2 million to local farmers, further tarnished the image of farmer-owned processing capacity in Wales.

A third type of producer experiment is worth mentioning here because it represents the most important politically inspired initiative to enhance the collective bargaining power of farmers in the supply chain. This is the Welsh Meat Company (WMC), which was originally designed to be a Wales-wide farmer-controlled livestock cooperative that would provide a 'one-stop' procurement and marketing service for beef and lamb. A prospectus was launched in spring 2000, with the aim of setting up a producer-owned cooperative which could guarantee the volume, quality, and continuity necessary to generate large-scale contracts to multiple retailers and the public sector. To be launched successfully, 1,000 farmers were needed to become members (some 6 per cent of total livestock farmers in Wales at the time) at a cost of just £250 each. However, the co-op failed to meet this target, attracting only 750 farmer subscriptions. This caused a good deal of political embarrassment for the Welsh Assembly Government, as such cooperative ventures were designed to be the delivery vehicles for its strategy for the development of the lamb and beef sectors. The venture was relaunched as a private company, rather than a cooperative, and it decided to move into value-adding activity, a sphere that had been specifically rejected in the original prospectus in deference to anxieties about the spectre of WQL. As well as designing higher value-added products, like the Welsh Lamb Sausage for example, the WMC has also undertaken a more robust branding exercise under the Celtic Pride brand of high-quality farm assured Welsh beef.

The WMC experiment represents the most important attempt to create a producer group that could strengthen the collective hand of farmers in dealing with the asymmetries of power in the red meat chain. Its failure to attract sufficient subscriptions to its original cooperative vision was symptomatic of a lack of trust in large-scale, top-down schemes, suggesting that farmers were far more prepared to identify with, and commit to, local producer group initiatives than to national ones. The fact that farmers are prepared to collaborate in local producer group schemes, such as Lleyn Beef for example, seems to prove that the problem is not collaboration per se, but the scale and the source of the initiative. If this is the case, the spectre of WQL may finally have been exorcized from the collective memory of the Welsh farming community.

If the commoditized nature of the product is the main cause of the dairy farmer's plight, the livestock farmer in Wales has reaped the benefits of a strategy which has successfully sought to 'decommoditize' Welsh lamb and meat products. At a generic level, the most important breakthrough came in 2003, when PGI status was secured for Welsh lamb and beef. It took years to achieve this certification victory due to objections from the abattoir sector, which argued that the animal should be slaughtered exclusively in Wales to be able to qualify for such status. Since Wales did not possess the processing capacity to slaughter everything it produced, this demand would have denied some livestock farmers the benefits of PGI status even though they technically produced the same high-quality product. To ensure that as many farmers as possible benefit from PGI status, the final PGI agreement applied to animals that were born, bred, and reared (but not necessarily slaughtered) in Wales. PGI status has undoubtedly helped Welsh livestock farmers to market their lamb and beef, especially in southern European countries, where such certification resonates with discerning private consumers and quality-conscious public-sector caterers such as the Roman school meals service.

At the more specialized end of the red meat sector, producers are reaping the benefits of the burgeoning demand for high-quality products that are not merely fully traceable but are also perceived to have additional quality features, such as Welsh Black beef and Welsh saltmarsh lamb. Organic red meat producers have sought to situate their product at this specialized end of the market. This specialized market, however, should not be seen as autonomous of, and separate from, the conventional red meat market, considering that over 80 per cent of Welsh organic meat is actually sold through the mainstream UK supermarkets. Although there has been a blossoming of novel ways of marketing organic red meat—through farmers' markets, mail order, farm gate sales, and the internet, for example—the fact remains that conventional retailers are the main gateway to the consumer for the vast majority of organic producers in Wales (Organic Centre Wales, 2004).

Within the organic meat sector, Graig Farm is one of the most successful enterprises, with its philosophy of 'telling the story behind the product' to consumers who are conscious of provenance and traceability. Sitting at the apex of a group of some 250 organic meat producers, Graig Farm produces its own and markets the group's organic meat products throughout the UK, using a wide array of marketing outlets, from the internet to the multiple retailer (see Ch. 3).

These innovative but small-scale producer groups seek to operate alongside, rather than in commercial opposition to, the larger retailer-led producer groups. In the brutal reality of the red meat supply chain in Wales, some 20,000 suppliers are facing as few as 6 major red meat buyers in the form of the multiple retailers. Notwithstanding the inherent asymmetries in this relationship, the case of Graig Farm seems to show that livestock farmers in Wales are coming to the conclusion that they have a viable future post-CAP reform if they become part of a well-managed producer group, maintain a quality product, and adopt a more commercial approach to their business, especially in terms of production costs (which continue to vary enormously) and their marketing strategies.

What differentiates the red meat and dairy sectors in Wales is as compelling as what they have in common. While they both suffer supply-chain asymmetries, with retailers increasingly calling the shots *vis-à-vis* the primary producer, the plight of the dairy farmer is far worse because of the nature of the product and the organization of the industry: as discussed above, milk is a commodity product in surplus and dairy farmers have failed to find collective solutions to their individual problems. This lack of organizational capacity will continue to take its toll in the dairy sector, where the conventional wisdom suggests that 'size is the key to efficiency with herd size increases critical to further cost reductions' (Colman and Harvey, 2004). In this scenario, the future holds little respite for the traditional family farm in Wales, where the average herd size is currently around 68, compared with a current UK average of 75 cows. The situation is in reality more complex than it seems; in fact, average herd sizes are much lower in the EU (30 in Ireland and as low as 18 in Spain for example), yet dairy farmers in these countries still manage to achieve higher prices than their UK counterparts (House of Commons, 2004).

In reality, smaller units can be viable, provided they organize themselves to find joint solutions to common problems and providing there is real demand for their product. These organizational and product quality issues have been at the forefront of a new agri-food strategy in Wales in recent years, which is designed essentially to help the country make the transition from one world of food to another.

Branding Wales: The Scope and Limits of the Agri-Food Strategy

The early years of the National Assembly for Wales coincided with a very difficult time from an agricultural standpoint: born in the midst of the BSE crisis, the Welsh Assembly government was hit by a second crisis, in 2001, in the shape of foot-and-mouth disease. Tragic as they were, these health scares diverted attention from the deeper problems of Welsh agriculture, particularly the commodity ghetto in which most producers were trapped, the culture of possessive individualism that stymied cooperation, and the supply-chain asymmetries that contributed to depress farm gate prices. These deep structural problems had been gestating for many years. Until the Assembly was created, however, there was little or no political incentive to challenge the status quo because, for the best part of the two preceding decades, the London-dominated Welsh Office was controlled by Conservative governments that were interested in managing, rather than transforming, this parlous state of affairs. Admittedly, a Welsh Food Strategy had been prepared in the mid-1990s, but it was marred by two fatal defects: first, it was designed in a top-down fashion and largely written by civil servants for the industry, which then felt no sense of ownership; second, it was addressed in the main to producers, with little cross-sectoral involvement of other players in the food chain.

What changed this moribund atmosphere was the emergence of a directly elected Welsh government genuinely committed to a new agenda—not just for agriculture, but for sustainable development in the broad sense of the term. A new generation of politicians, some of whom had a strong track record of environmental activism, helped to move away from agriculture as a narrow sectoral interest and to develop a more integrated vision that connects agriculture to sustainable development. One immediate sign of this political sea-change was the decision to drop agriculture from the title of the new portfolio, which was quite consciously christened Environment, Planning, and Countryside. The most important sign of the new political era was section 121 of the Government of Wales Act 1998, which made it a legal duty for the Assembly to promote sustainable development (WAG,

2000). For the first time, a government of any jurisdiction anywhere in the EU was legally required, by its political constitution, to promote sustainable development. By committing itself to a sustainable, multifunctional agriculture, the Assembly raises issues that resonate around the world, especially in countries that aspire to put sustainable development principles into practice.

The task of promoting a new strategy fell to the Agri-Food Partnership, a cross-sectoral body formed in 1999 to provide focus and a more 'joined-up' approach to the development of the agri-food sector in Wales. The membership of the Partnership consists of a combination of public and private sector interests from the whole supply chain. Three sector groups were created to develop action plans in the Red Meat, Dairy, and Organic Sectors and two development groups were set up to address cross-cutting issues. In 2002 two additional sector groups were established for Fisheries and Aquaculture and for Horticulture. The whole structure is shown in Fig. 6.1.

This Partnership represents a major advance over the earlier Welsh Food Strategy in two vital respects: in membership terms, it is less biased to the

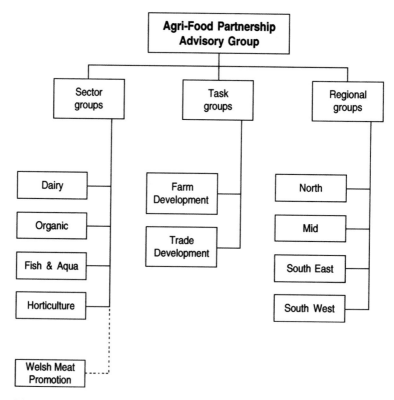

Fig. 6.1. The structure of the Agri-Food Partnership in Wales
Source: DTZ (2004).

supply side and draws on all parts of the agri-food chain; in management terms, it is run by and for its members, with the result that there is a strong sense of commitment to, and ownership of, the strategy. The task of animating and orchestrating the Partnership, especially in its early stages, was performed by the Welsh Development Agency, which had the foresight to create a new Food Directorate to give more focused support to a sector that had been hitherto neglected. At the heart of the new Agri-Food Strategy are four cross-cutting strategic goals, namely: to improve market focus; to improve supply-chain linkages; to improve the performance of processors; and to improve the performance of the primary producer. In view of the Assembly's legal duty to promote sustainability, these strategic goals are embedded in a wider vision that aims to make the connections with other policy priorities, as shown in Fig. 6.2.

The Agri-Food Strategy draws its vision from *Farming for the Future*, a seminal report which offered a brutally frank assessment of the nature of the problems and of the action required to deal with them. Published in 2001, this blueprint was the brainchild of a group of experts, the Farming Futures Group, drawn together by the Minister to provide advice on 'a new vision for the future of Welsh agriculture' (WAG, 2001). As the excerpts included in Box 4 show, the Welsh strategy puts the social dimension of sustainability ahead of all others.

Farming for the Future is arguably the most explicit attempt in the UK to forge a direct connection between the quality of food production and the quality of life of the small food producer, a link which is deemed to be crucial to the preservation of the family farm in Wales, the rural repository of Welsh-speaking culture. The *social* dimension of sustainable development in Wales expresses itself most clearly in the Assembly's strong support for the family farm, a commitment that is wholly absent from the Curry report, which articulates the official vision of the future of food and farming in England (Morgan, 2002). This political commitment to the family farm in Wales also helps us to understand why Wales diverged so radically from England in the way it elected to implement CAP reform, preferring the historic system to the area system because the former gave small producers the best deal under the new Single Farm Payment.

If *Farming for the Future* is unashamedly supportive of the social cause of the family farm, it is also unambiguously clear about the key economic challenge facing agriculture: namely, that collaboration is the route to a quality-driven agri-food sector producing higher value-added, branded products for more specialized markets. Since branding and traceability are clearly crucial to this endeavour, the Partnership has set a high premium on differentiating Welsh food brands by using national (i.e. Welsh), regional and local logos, as well as the exploitation of PGI/PDO status.

One of the distinctive features of this new Agri-Food Strategy is that its reach extends way beyond the realm of the primary producer, the usual

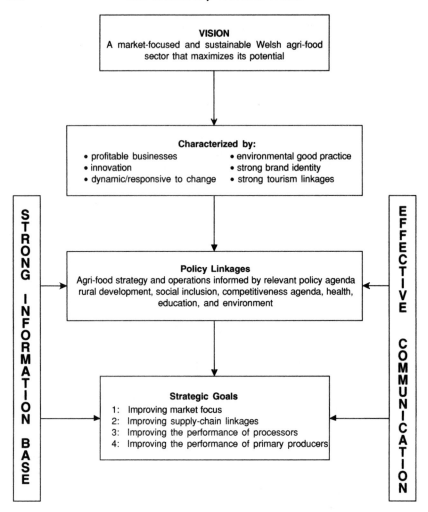

Fig. 6.2. The Agri-Food Strategy in outline
Source: DTZ (2004).

supply-side target of such strategies. The Partnership is acutely conscious of the limits of a one-sided, producer-driven approach, which was the major shortcoming of its first Organic Action Plan. The rapid increase in organic supply led to severe marketing problems in certain product segments, under-lining the need for a more judicious mix of demand 'pull' and supply 'push' measures, the approach adopted in the Second Organic Action Plan (Agri-Food Partnership, 2004).

Stimulating the demand-side of the market for quality food products is, in fact, a central thread running through the whole strategy. Local Food Festivals, for example, have been energetically promoted to help consumers

Box 4.   A new direction for farming in Wales

'The National Assembly Government's objective is to maintain viable and balanced communities in rural Wales. Helping agriculture to adapt has an important part to play in this. There are some who advocate a polarisation in agriculture, so that food would be produced intensively in lowland areas while upland areas are allowed to revert to wilderness, or managed primarily for tourism. The National Assembly Government rejects such arguments. The social implications for rural communities would be very damaging. There would be dramatic changes in the character of the Welsh countryside. Consumers would see still more intensification in food production systems. The objective should be to promote agriculture which is sustainable economically, environmentally and socially in all areas. What Wales therefore needs is an agriculture which delivers the following outputs:

- Safe, healthy food and non-food products, produced with high standards of care for the environment and animal welfare and targeted much more closely on market opportunities to give farming families a better return

- A countryside which is visually attractive and rich in its biodiversity, archaeology, history and culture, not only for its own sake but for people's enjoyment and to help support tourism

- Distinctive local food products as the basis for a cuisine which helps promote tourism and which, through all the above, contributes to a positive image for Wales in the world.

The principles of sustainable development, in all its aspects, are therefore at the heart of this strategy. Organic farming is one way of farming sustainably and has a great deal to offer Wales. It epitomises much of what the new strategy means. The National Assembly has led the way in Europe in acting to prevent cross-pollination from Genetically Modified (GM) plantings, to safeguard organic production.... Models of conventional farming which embody many of the principles of sustainability also exist and the National Assembly Government will work to promote sustainability through both organic and conventional farming.'

Source: WAG (2001).

rediscover the links between products and places. The annual True Taste Food and Drink Awards, first launched in 2000, have helped to showcase a wide array of produce, with the winning entries going on to secure contracts to supply such premier London clients as the world-famous Savoy. To this end, international food competitions are contested more seriously than ever before, with some encouraging results. Welsh saltmarsh lamb, for example, secured the top prize in a competition adjudicated by leading Parisian chefs, while Welsh Black won the best beef competition in London in 2003 (Clarke, 2003). To complement the fine restaurant side of the market, the public procurement market has also been targeted for special attention. Until recently, this segment had strangely been ignored despite the fact that schools, hospitals, care homes, and the like collectively spend some £60 million p. a. on purchasing food. Tapping the potential of this public catering market requires a cultural change among both buyers and suppliers because the former are used to foraging in low-cost and low-quality supply chains, while the latter have never developed the specialized skills to secure public-sector catering contracts. This scenario is beginning to change because the Welsh Assembly Government has made a strenuous effort to promote a more creative and more concerted public procurement policy, which aims at securing a triple dividend: a health dividend from more nutritious food; an economic dividend from more localized markets for local producers; and an environmental dividend from lower food miles (Morgan and Morley, 2002).

Calibrating supply and demand is not the only distinctive feature of the new Agri-Food Strategy. The Farming Connect scheme, which provides a comprehensive range of business support to the farming community in Wales, has won accolades for its innovative approach, not least from Lord Haskins, the UK government adviser, who claims it is one of the best examples of integrated rural development policy in Europe (C. Jones, 2004). At the core of Farming Connect, a scheme originally launched in 2001, is the Farm Business Development Plan, a free health check for each individual farm, offered in conjunction with a wide array of related services, including subsidized training opportunities, technical advice, capital grants to assist farm investment, and access to a Wales-wide network of demon-stration farms to facilitate knowledge transfer from one's own peers. In many ways, Farming Connect was ahead of its time because it anticipated by several years the Farm Advisory System, one of the institutional innovations of the new CAP reform package, which is designed to help farmers develop their farms as commercial undertakings.

In essence, then, Farming Connect is a targeted business support scheme designed to build the supply-side capacity, while cognate market-making measures are intended to foster the demand-side, a calibrated strategy to enhance the prospects for quality food. Laudable as it is, however, the new Agri-Food Strategy faces enormous obstacles. The fact that *Farming for the Future* contains a staggering fifty-two action points speaks volumes for the scale of the challenge involved in making a successful transition from one world of food to another.

For all the achievements of the Agri-Food Partnership, a recent review of its strategy (DTZ, 2004) suggested that the de-bundling of the agri-food sector into a series of development groups carries costs as well as benefits. While the sector groups undoubtedly provide focus and drive, this format can also lead to 'an introverted and self-interested perspective where all the energies are directed at supporting your own group with scant attention paid to the objectives and efforts of others'. This 'silo' approach ran counter to the central aim of the Agri-Food Partnership, which was to promote an integrated or 'joined-up' approach to the development of the whole sector. The key weaknesses of the strategy to date were deemed to be with the Partnership's overarching Advisory Group, which was found to be 'punching below its weight' in at least three respects: it was not doing enough to challenge the status quo; it could do more to promote good practice across all the groups; and it needed to be more vigilant in monitoring the delivery of the strategy. With so many action points, each carrying its own transaction costs, it is perhaps inevitable that a strategy that has erred on the side of ambition will fall short of some of its targets.

Devolution and Divergence: Beyond the Commodity World?

The 'differentiated countryside' was a fact in the nations and regions of the UK long before the advent of democratic devolution in 1999 (Murdoch et al., 2003). What devolution did, in political terms, was to give voice to these differentiated systems, enabling the devolved administrations in Scotland and Wales to pursue policies that were more attuned to their own specific circumstances. Despite its modest powers, the National Assembly has provided a major impetus to the design and delivery of a distinctly *Welsh* agri-food strategy. This was something of an achievement in itself because, prior to the Assembly, there was not a Welsh sector to speak of in any meaningful sense of the term; even the statistical profiles of the sector referred to the broader geopolitical amalgam of 'England and Wales', an entity that threatened to denude Wales not merely of a sectoral identity in agriculture but of its national identity (Morgan and Mungham, 2000).

The implementation of the 2003 CAP reform package in Wales was interpreted as a major 'victory for devolution' because for the first time the principle of subsidiarity was applied *within* a member state (Dube, 2004). In this respect, the devolved administrations in the UK secured a historic breakthrough at the 2003 CAP negotiations in Luxembourg by winning the right to implement the reform according to their own political preferences. Although the UK government wanted the new Single Farm Payment (SFP) to be a flat-rate area payment based on acreage, the system that was actually selected for England, the National Assembly chose the historic payment system, which means that the SFP in Wales is calculated on the average claims farmers made during the years from 2000 to 2002. The rationale for choosing the historic route in Wales was chillingly simple according to Carwyn Jones, the Environment Minister (quoted in Dube, 2004):

we looked at all the options and this is the least destructive outcome. The area system would have distributed at least half the money away from small farms to large farms and a hybrid system would have done the same. The bigger the farm, the more money it would get and the family farm would have been out on its own.

Not surprisingly, the historic payment decision received a rapturous welcome from the farming unions, with the NFU president in Wales going so far as to say that this was 'probably the most important decision on agriculture that the Assembly has ever made' (ibid.). Although the Assembly perceives the English option for an area-based system as favouring a productivist model, in which large farms and agri-businesses are the basic units of production, critics of the Welsh decision claim there is an alternative, non-productivist argument for having an area-based payment system. In this scenario, a flat-rate area-based payment system, combined with robust rural development policies under Pillar Two of the CAP, would create better opportunities because it would favour a more extensive agriculture with lower stocking levels—compared with a historic payment system that per-petuates past levels of output and threatens to set the status quo in aspic (Midmore, 2004).

While the Minister conceded that there were potential dangers with the historic system, which could freeze the status quo, he argued that the new Agri-Food Strategy would temper the pull of the past and the historic pay-ment system would help to alleviate the plight of the small primary producer in Wales. Whatever the shortcomings of the CAP decision, the fact remains that, without a devolved decision-making capacity, Wales would have been obliged to implement an area-based system designed for a very different type of agriculture and a different set of social and cultural values.

An even more dramatic example of divergence between Wales and Eng-land is the radically different positions that have been adopted on GMOs by the administrations in Cardiff and London, two of the most extreme ends of the governmental spectrum in the whole of the EU. On the one side of this debate, Tony Blair has declared that he is personally committed to exploiting GM technology to help to maintain the UK's position as a leading research centre in the life sciences. On the other side, the National Assembly voted to adopt 'the most legally restrictive policy' on GM crops, a move prompted by a desire to protect the flourishing but vulnerable Welsh organic sector, which represents a key component of the new agri-food trajectory in Wales. In 2001 the Assembly became the first administration in the UK to have in place legally enforceable separation distances between GM and non-GM maize, a political innovation that triggered a new debate in the EU on the role of GM crops and how they might interact with both organic and conventional crops. More significantly, in 2003 Wales joined Upper Austria and Tuscany, along with seven other European regions, to form a network of regions that aspire to remain GMO-free. Among other things, this network invites the EU to agree that 'European regions could define their own territory or part of it as a GMO-FREE zone ... without these decisions being considered as an infringement of the free movement of goods principle' (Network of GMO-Free Regions, 2003).

The Welsh Environment Minister prefers to describe the Assembly's policy as 'the most legally restrictive' within EU and UK law, rather than as a GM-free policy per se, since he cannot guarantee that the latter has not already been breached. He also insists on the need for international action because, as he argues, 'GM crops do not respect national boundaries, which is why it is better to take decisions on anything pertaining to the environment on a wider, broader level—at European or world level if possible—rather than concentrating on finding a purely Welsh or British solution' (C. Jones, 2003).

In conclusion, Wales has capitalized on the new opportunities for regional distinctiveness offered by devolution and CAP reform to shape its own quality-based world of food. Indeed, the establishment of the Agri-Food Partnership has brought together a number of different private- and public-sector actors operating across the whole supply chain 'to provide focus and a "joined-up" approach for the development of the agrifood industry in Wales' (WDA, 2004: 1). As in Tuscany, this approach has been supported by a new discourse that, as our analysis of *Farming for the Future* has shown, is promoting the distinctiveness of Welsh rural life and small-scale farming through a recognized need to compete on the basis of social, economic, and environmental 'quality' food criteria, rather than on quantity and price. In this context, agriculture and agri-food are re-emphasized and repositioned for their contribution to achieving environmental and socio-economic sustainability (Marsden and Sonnino, 2005).

The question now is whether the new discourse can deliver a genuinely sustainable trajectory of development, which in the context of agri-food means effecting a genuine break with the commodity world in which Wales has been ghettoed for the best part of a century. In this respect, there are encouraging signs at the production end of the food chain. As our Graig Farm network case study shows (Ch. 3), producers in Wales no longer perceive the status quo as a viable option for the future and there is a growing belief that the way forward lies in collaborating among themselves and with others downstream in the food chain.

However, to envisage a future that is more than a continuation of the past—a future in which Welsh produce becomes associated with high-quality branded products—intervention is needed also at the consumption end of the food chain. In a context where there is no tradition of localized consumer culture as there is, for example, in Tuscany, political action is needed not just to create a friendly regulatory structure, but also, and perhaps most importantly, to create knowledgeable consumers who can sustain a market for a high-quality Welsh world of food.

7

Beyond the Placeless Foodscape: Place, Power, and Provenance

The foregoing chapters bring us to the point where we can directly address the three themes that constitute the subtitle of the book—namely place, power, and provenance. Reflecting the binary thinking that pervades the agri-food literature—global versus local, embedded versus disembedded, conventional versus alternative, quantity versus quality, and so forth—these themes tend to be treated in a highly compartimentalized fashion, with *place* and *provenance* being the preserve of the alternative food literature, while *power* seems to be the proper object of analysis in the conventional food literature. This binary conceptual tradition has the effect of segmenting the food sector into unduly rigid and path-dependent worlds of production. It could even lead to the (erroneous) conclusion that the conventional food chain is inextricably tied to a particular world of production, invariably the Industrial World, while alternative food chains are embedded in, and tethered to, the Interpersonal World.

To overcome this unwarranted division of labour, we propose to examine the roles of place, provenance, and power in both the conventional food chain and the ecological food chain. However, we also want to suggest that the borders between these worlds are more porous and much less static than the worlds of production literature sometimes implies, leaving open the possibility that firms and regions can move from one world to another.

Each world of production may have its own nuanced regulatory environment, where a specific mix of rules, regulations, and quality conventions defines its distinctive milieu, but all worlds are subject to some meta-regulatory trends that are emerging in the global food sector, two of which have the potential to induce significant changes. For the sake of simplicity, we shall refer to these meta-regulatory trends as the *new moral economy* on the one hand and the *neo-liberal economy* on the other.

Taken in isolation, these regulatory trends could trigger very different trajectories of development, with major implications for place, power, and provenance in the food chain, because the former involves reregulating the food sector, while the latter aims to deregulate it. In reality, of course, these meta-regulatory trends will evolve not in isolation but in tandem, creating a whole series of tensions and conflicts in the multilevel governance system, from the global level of the WTO to the local level of municipal government. To illustrate these tensions and conflicts, the following section examines the new moral economy of food as a prelude to addressing the themes of place, power, and provenance in different food chains.

The New Moral Economy of Food

The concept of the moral economy has resurfaced in recent years, partly as a response to the excessive utilitarianism of mainstream economics and partly as a vehicle for academics and activists to address the normative issues that they consider to be *intrinsically* significant (such as health, education, and well-being), rather than merely instrumentally significant (such as money). According to Sayer, a prominent social theorist in this field, the concept of the moral economy can be defined in the following way (Sayer, 2000):

the moral economy embodies norms and sentiments regarding the responsibilities and rights of individuals and institutions with respect to others. These norms and sentiments go beyond matters of justice and equality, to conceptions of the good, for example regarding needs and the ends of economic activity. They might also be extended further to include the treatment of the environment. The term moral economy has usually been applied to societies in which there are few or no markets, hence no competition and law of value, and in which economic activity is governed by norms regarding what people's work responsibilities are, what and how much they are allowed to consume, who they are responsible for, beholden to and dependent on. However, moral norms … are also present and influential in advanced capitalist societies, though they tend to be overlooked by political economy, radical or otherwise. They exist both within the formal, money economy and outside, particularly in the household economy. While the norms may be considered part of a moral order, both the norms themselves and the associated behaviour are invariably influenced by networks of power and considerations of cost and risk.

Although this is offered as a generic definition of the moral economy, it turns out to be highly germane to contemporary debates in and around the global agri-food system. From the ecological standpoint of this book, one of the great merits of Sayer's conception is that it helps to overcome 'the dualistic separation of nature and society', which has been one of the abiding weaknesses of the mainstream agri-food literature (Goodman, 1999). Another advantage is that it provides an ideal context in which to frame the discussion of health, well-being, fair trade, and development, the quintessential dimensions of the new moral economy of agri-food. Sayer's moral economy clearly has strong affinities with 'green political economy', which challenges the way that societies value nature and, in particular, 'challenges the valuation of nature purely on the basis of individual preferences expressed through market choices or cost–benefit analyses instead of through political and ethical argument' (Sayer, 2000).

Both the above conceptions agree that conventional political economy tends to overlook the moral claims of society or nature because of its fixation with unrestrained economic power in terms of narrow self-interest. Paradoxically, though, only a moral economy perspective can help us to understand why the most powerful economic actors, namely the multinationals, feel constrained to exercise their power in a naked and unencumbered manner in their dealings with weaker interlocutors. We are not of course suggesting that these corporate leviathans recoil from striking the most advantageous deals possible, but that they do so, to use Sayer's terms, in the context of prevailing moral norms. As we shall see, multinational retailers clearly have the power, at least in theory, to secure much lower prices from their developing country suppliers than they enjoy at present. The fact that they refrain from doing so in practice is a testament, however modest, to the tempering (and civilizing) effects of prevailing moral norms that have been established through a messy combination of multilateral political agreements, NGO pressure, and the moral sentiments of affluent consumers at home. To explore the new moral economy of agri-food in more detail, in the following sections we will examine its two key dimensions.

The Moral Economy of Health

Despite its celebrated lobbying power, the conventional food industry in Europe and the US manifestly failed to anticipate what turned out to be one of the biggest challenges it has ever faced: namely, 'the global epidemic of obesity' (World Health Organization, 1998). The fact that a poor diet is a risk factor in a whole series of non-communicable diseases—such as cancer, coronary heart disease, obesity, and diabetes for example—had been well known, and widely accepted, for decades. However, it was not until the early years of the twenty-first century that one of these diseases—obesity—caught the political imagination of governments in developed and developing countries alike. Although just one of a number of diet-related diseases, obesity began to assume a totemic status in the public mind largely because it was so visible, because it was rapidly developing among children, and because new estimates of the human and financial costs were beginning to appear more regularly around the world. In the advanced OECD countries, obesity is most pronounced in the lower socio-economic classes, especially among women and in rural communities. In developing countries, by contrast, obesity tends to be more common in the higher socio-economic classes, reflecting the historical situation in the advanced countries, where 'only the rich could afford to get fat' (Lang and Heasman, 2004).

The advent and diffusion of cheap, highly processed food, high in fat, sugar, and salt, is widely believed to be one of the main causes of the global epidemic of obesity. The conventional food industry tries to absolve itself of any responsibility for obesity, arguing instead that sedentary lifestyles are the root cause of the problem, when in fact this is caused by a cocktail of factors, poor diet and inadequate physical exercise being two of the principal ones. It is debatable whether the 'moral panic' about obesity genuinely concerned the multinational food companies. What really goaded them into action were two particular threats: the threat of anti-obesity litigation from obese consumers and the threat to their share prices from a nervous investment community.

Anti-obesity litigation began with the landmark *Pelman* v. *McDonald's* case in the US, where two New York teenagers filed a suit against McDonald's for making them fat and, in particular, for allegedly concealing the health risks of Chicken McNuggets. Although the case was dismissed, the judge's verdict was very damning for the fast-food giant. Among other things, he said that a combination of additives and a high fat content made Chicken McNuggets more than just fried chicken, he admitted that reasonable consumers might not know this, and he called the product a 'McFrankenstein creation' (Buckley, 2003). An anti-obesity litigation industry sprung up in the wake of this landmark case, led by the attorneys-general, many of whom are focusing on the cost to their states of maintaining Medicaid payments to low-income patients with obesity-related diseases. These lawsuits remove the obese person from the litigation and substitute financial loss to the state's finances (Grant, 2005). The implications of these trends were not lost on the food industry, particularly the suggestion that fast food may be 'addictive', creating a dangerous association between 'big food' and 'big tobacco'. New scientific findings from Princeton University suggest that high fat and high sugar foods create biochemical reactions similar to those seen in people addicted to tobacco. If true, these findings could undermine the key legal defence of the fast-food industry—that consumers are wholly responsible for their weight. Far from being a matter of gluttony or fecklessness, obesity seems to be the result of vulnerable genes in a hostile environment (Shell, 2002).

The international investment community quickly recognized the implications of this new moral consciousness about food. One of the first to offer an equity analysis was UBS Warburg, which unnerved the food industry by suggesting that the risks associated with obesity 'have not yet been factored into share prices' (UBS Warburg, 2002). Even more dire warnings were issued the following year by equity researchers at JP Morgan, who forecast that manufacturers of fast food, soft drinks, and snacks faced an ever growing risk of tougher regulation, especially in Europe, in their key growth products. The JP Morgan analysts even drew up a new set of metrics for judging food companies based on the product portfolios most exposed to the 'obesity risk'. In terms of the percentage of total revenue derived from 'not so healthy' food, the analysts found that the league table was headed by Hershey, with a 95 per cent exposure, followed by Cadbury (88 per cent), Coca-Cola (76 per cent), PepsiCo (73 per cent), and Kraft (51 per cent) (Lang and

Heasman, 2004). These firms were also heavily implicated in what many health and consumer groups consider to be the most egregious practice of all: namely, the marketing of unhealthy food and drink products to children, mostly in the form of child-directed TV advertising. A recent report from the International Association of Consumer Food Organizations, called *Broadcasting Bad Health: Why Food Marketing to Children Needs to Be Controlled*, came to the following conclusions:

- the food industry's global advertising budget is $40 billion, a figure greater than the GDP of 70 per cent of the world's nations;
- for every dollar spent by the WHO on preventing the diseases caused by western-style diets, more than $500 is spent by the food industry promoting these diets;
- in industrialized countries, food advertising accounts for around half of all advertising broadcast during children's TV viewing times. Three-quarters of such food advertisements promote high-calorie, low-nutrient foods;
- for countries with transitional economies (such as those of eastern Europe), typically 60 per cent of foreign direct investment in food production is for sugar, confectionery, and soft drinks; for every $100 invested in fruit and vegetable production, over $1,000 is being invested in soft drinks and confectionery;
- over half the world's population lives in less-industrialized countries such as Russia, China, and India and they are now suffering a rising tide of diet-related diseases as food companies export their products and their advertising practices.   (Quoted in Lang and Heasman, 2004.)

To counter its critics, the food industry's response consists of two well-rehearsed arguments: on the health front, it maintains that there is no such thing as unhealthy food only 'unhealthy diets' and, on the ideological front, it argues that food choice is a private not a public matter. Governments that try to promote a more nutritious food policy are immediately accused by the food industry of behaving in a 'totalitarian' manner because they make an unwarranted intrusion into the private realm of the individual and treat food choice as an issue of the public realm (Nestle, 2002).

Yet, if the burgeoning health cost of diet-related disease is not a legitimate issue for the public realm, it is difficult to imagine what is. Ever since the US Surgeon-General issued 'a call to action' to combat escalating obesity levels in 2000, there has been growing political concern that diet-related diseases could, if unchecked, bankrupt the public health service in the worst-affected countries. By common consent, the US is the worst-affected country, with the healthcare cost of obesity-related illness reaching $117 billion in 2001 and showing no sign of abating (Buckley, 2003). One of the key targets of obesity litigation in the US is constituted by education boards, which have allowed fast-food companies to colonize school vending machines, in some cases offering exclusive rights (so-called 'pouring rights') to soft drink companies that share with the school the proceeds from increased soda consumption (Nestle, 2002).

This very brief overview of the obesity epidemic is just one aspect of the escalating human and financial costs of diet-related disease. If, as Sayer suggests, the study of moral economy focuses on the norms regarding the rights and responsibilities of individuals and institutions with respect to others in production, distribution, exchange, and consumption, then it is clearly appropriate to speak of the emergence of a new moral economy of diet-related disease. At the heart of this new moral economy is a highly contested debate about responsibility. To what extent should governments be responsible for the social environment of food choice? Should the food industry desist from selling foods of low nutritional value? How culpable is the individual when there is little nutritional information available from producers and little knowledge on the part of consumers? And, most important of all, is it morally acceptable to treat children as discerning and responsible consumers to whom products of low nutritional value can be marketed on a routine basis?

The fact that these questions are now posed more forcibly than ever suggests that a new moral consciousness is beginning to take shape around *conventional* food consumption, so much so that ethical considerations are no longer confined (if indeed they ever were) to alternative or exotic forms of consumption (Sassatelli, 2004). The growing moral questioning of conventional food, or 'ordinary consumption' (Gronow and Warde, 2001), is both a cause and a consequence of the erosion of trust in conventional products, which suggests that 'ordinary' consumers are engaged in requalifying foodstuffs. This requalification challenges the habitual, taken-for-granted assumptions about food quality that have prevailed in the post-war era, a process in which consumers begin to '*distance* themselves from food goods in order that they might *reconnect* in new ways' (Murdoch and Miele, 2004).

One of the potential virtues of the obesity epidemic is that consumers might become more aware of the hidden costs of so-called cheap food. This means that the food might seem cheap at the point of sale, 'but the citizen/taxpayer pays additional costs for healthcare later' (Lang and Rayner, 2003). In terms of the capacity to affect public health, poor diets are more common, and far more of a threat, than episodic food scares.

Containing and reducing the burgeoning costs of diet-related disease will require action at a number of different levels. For example, governments will need to play a more robust role in reforming the social environment of food choice to make healthier options more accessible and affordable. For their part, food companies will have to respond to moral pressure from consumers and political pressure from governments to develop healthier product lines. And if consumers are to assume more responsibility for their health, as they are increasingly enjoined, they will require better information about their food, particularly with respect to its nutritional value, its ingredients, its country of origin, and its methods of production. Whereas, in theory, food labelling policy should help to empower consumers here, all the evidence suggests that it has a bewildering, rather than an empowering, effect, as we shall see later.

The Moral Economy of Fair Trade

Another important dimension of the new moral economy is the debate around fair trade for developing countries. The creation of a fair trade system through multilateral agreement at the WTO needs to be distinguished from Fairtrade (FT) labelled products. One might think that there would be less need of an FT *label* if we had a fair trade *system*. In this respect, the WTO claims to be designing the rules of a new, and ostensibly fairer world trade system. Indeed, this is one of the formal aims of the Doha Development Round, which was called as such to signal to developing countries, a majority of the WTO's members, that their interests would be paramount in the making of a new multilateral trade agreement. Formally, then, the Doha Round is seeking to introduce a stronger moral economy perspective into the world trade system by rewriting the rules of the game in such a way that unequal countries are not treated equally, a sure way to reproduce inequality under the guise of equality.

In our discussion of the Doha Round in Ch. 2 we underlined the significance of the Development Box, a concept that embraced the 'non-trade' concerns that are vitally important to any conception of fair trade. Under pressure from developed countries, and especially from big agri-exporting countries such as the US, the Development Box idea was watered down into a less radical version of Special and Differential Treatment (SDT). The relationship between trade and development is central to the negotiations on SDT, a concept that stretches back to the earliest years of the GATT. With the advent of the WTO, however, the SDT provisions were weakened in the Uruguay Round. This rendered them less effective as a mechanism for promoting indigenous development and fair trade. Developing countries agreed to this change at the Uruguay Round in anticipation of the benefits from increased market access in agriculture, textiles, and clothing, along with full implementation of the SDT provisions. As we saw in Ch. 2, however, most of these benefits failed to materialize in the years following the Uruguay Round. Understandably cynical and frustrated, developing countries have since decided to refocus their efforts on the SDT provisions and to make them more 'precise, effective and operational' (ICTSD, 2004).

To this end, developing countries have made two specific SDT proposals. The first is for a *Special Products* category which would allow them to designate certain crops—those vital to food security, livelihoods, and rural development—as exempt from tariff cuts. The introduction of this idea into the Doha negotiating framework signals a major change in WTO thinking because it means that the Geneva-based body has finally accepted that not all crops are equal. The second SDT proposal is for a *Special Safeguard Mechanism*, which would allow poor countries to increase tariffs temporarily in the face of fluctuating import prices or volumes (Oxfam, 2005). SDT provisions might look like dry and arcane technicalities, but they constitute the main political battleground on which the campaign for a fairer trade system will be won or lost.

The US is leading the campaign for a narrow definition of SDT, arguing that it should apply to a restricted range of farm products, essentially those produced by subsistence farmers. 'Food security', said a US trade official, 'is often better served by opening markets to high-quality, low-cost produce than favouring domestic producers' (Beattie, 2005). The US is opposed to a broad SDT agreement principally because this would encourage developing countries to provide for their own food security, thereby reducing the export markets for US agri-business. In a remarkably candid statement, the USDA (2001) said that its agri-food strategy for the new century was predicated on a fusion of the local and the global, because

domestic demand alone is no longer sufficient to absorb what American farmers can produce. Demand by well-fed Americans grows slowly, with population growth. The promise of new, much faster-growing markets lies overseas. ... As a result, the US must consider its farm policy in an international setting, helping farmers stay competitive while pressing for unfettered access to global markets.

A recent Oxfam analysis illustrates the vested interests at stake in the SDT dispute with reference to rice, the staple food for half the world's population (Oxfam, 2005). Rice is more than an ordinary crop, it is a way of life, the means through which the poor pay for the manifold needs of their households. It is also a crop that is largely cultivated and processed by women, whose earnings are crucial to poverty reduction in rural areas. However, the US has developed a large interest in rice and it has become the world's third largest exporter of it—even though its rice costs twice as much to grow as it does in Thailand and Vietnam, the world's top rice exporters. This situation is made possible because of lavish state subsidies: in 2003, for example, the US government ploughed $1.3 billion into rice sector subsidies, supporting farmers to grow a crop that cost them $1.8 billion to produce. Between 2000 and 2003 it cost on average $415 to grow and mill one tonne of white rice in the US, but that was dumped on export markets for $274 per tonne, 34 per cent below its true cost of production. As Oxfam states, the 'real winner from this combination of subsidy bonanza in the US and rapid trade liberalization in developing countries is US agri-business' (ibid.).

To be remembered as a genuine development round, the Doha trade round will have to temper the global ambitions of US agri-business and design a fairer set of world trade rules. A new and fairer WTO Agreement on Agriculture could begin quite simply, in Oxfam's view, with the WTO clearly committing itself to its own negotiating text, which says that 'developing country members should be able to pursue agricultural policies that are supportive of their development goals, poverty reduction strategies, food security, and livelihood concerns' (ibid.). Giving developing countries more autonomy over their domestic policy space, helping them to access developed country markets, and ending export dumping would herald a genuine shift to a fairer trade system.

In the absence of a fair trade *system*, the Fairtrade (FT) *label* helped to fill the vacuum. FT labelling was created in the Netherlands in the 1980s and FT products have managed to establish an identity for themselves in a relatively short space of time. A major organizational innovation came in 1997, when the Fairtrade Labelling Organization was formed to provide an independent international body to oversee the certification of FT products. Over 70 per cent of FT goods consist of food and drink products, mainly in the form of coffee, cocoa, bananas, and sugar. FT goods are now sold in seventeen countries and these are sourced from 360 production groups in forty countries, representing a total of 4.5 million growers and their families. Sales of FT products in the eighteen countries in which they are licensed are growing by roughly 20 per cent a year, and some of these products, such as coffee and bananas for example, are becoming mainstream products in conventional supermarkets.

The central goal of the FT label is to change international commercial relations in such a way that disadvantaged producers can increase their control over their own future by earning a fair return for their work and by having better working conditions. In short, the FT label helps to 'humanize' trade relations by making the 'producer–consumer chain as short as possible so that consumers become aware of the culture, identity, and conditions in which producers live' (Raynolds, 2003). The FT transaction is therefore much more than a conventional commercial transaction. Guillermo Vargas, a member of the Costa Rican Fairtrade coffee cooperative, emphasized this point during his European tour in 2002, by saying that 'when you buy Fairtrade you are supporting our democracy' (Morgan and Morley, 2002).

A more localized form of solidarity was injected into the FT campaign in 2000, when Garstang, a small town in the north of England, declared itself to be 'the world's first Fairtrade Town'. The move earned the town such positive publicity that it stimulated other areas to follow its example, with the result that more than 220 towns, cities, and smaller-scale zones had acquired Fairtrade status by early 2005. Five conditions have to be fulfilled before an area—be it a town, city, or zone—qualifies for Fairtrade status:

- the local council must pass a resolution supporting Fairtrade and serve FT tea and coffee at all its meetings;
- a range of FT products must be readily available in the area's shops, with targets set in relation to population;

- FT products must be used in local workplaces and community organizations, especially schools;
- the council must attract support from the local civil society for the FT campaign;
- an FT steering committee must be formed to sustain the commitment to Fairtrade goods.   (Fairtrade Foundation, 2005)

One of the most remarkable aspects of the Garstang story is that the campaign to highlight the plight of primary producers abroad resonated with the campaign to help primary producers at home (see Box 5).

Although the FT label can never be a substitute for a fairer trade system, the rationale for FT-labelled products will not disappear if and when the WTO achieves a more equitable world trade regime. This is because the FT label seeks to address *development* issues—such as labour standards, ecological practices, and participative governance structures, for example—that might not be part of a new *trade* system.

One of the common threads running through these two dimensions of the new moral economy—human health and fair trade—is that they are led not by governments but by new social movements, notably consumer associations, development agencies, and environmental groups (Murdoch and Miele, 2004). In the case of health, we noted the erosion of public trust in the conventional food system, especially in Europe, with the result that consumers are beginning to place a higher premium on such attributes as the provenance of their food. This suggests that, contrary to the tacit assumptions of the agri-food literature, ethical and quality considerations are no longer confined to the alternative food sector.

If the health example signals a new relational reflexivity, with consumers concerned about their own well-being as well as that of their children, the fair trade case illustrates a very different sentiment, namely, a regard for the unknown and distant 'other'. Although the moral economy literature from Smith to Sayer notes that moral sentiments have a tendency to decline with distance, the diffusion of FT goods and the proliferation of Fairtrade Towns suggests that consumer–producer linkages can be a mechanism, however modest, to affirm international solidarity between a rich North and a poor South. Although the FT label is just one modest component of a fair trade system, it highlights issues that are pivotal to the new moral economy of international trade and development.

The growing concerns for human health and fair trade imply the need for a more robust regulatory approach to the agri-food sector, a very different prognosis from the neo-liberal concern for liberalization and deregulation. In both health and fair trade, powerful agri-business interests are demanding a lighter regulatory regime, creating new tensions within and between the conventional and ecological food systems, as we will see in the following sections.

Box 5. Garstang: the world's first Fairtrade Town

Garstang is a historic Lancashire market town with a population of just over 4,000 people. At a town public meeting in April 2000, the people of Garstang voted unanimously for Garstang to become 'the world's first Fairtrade Town'. The driving force behind this declaration was Garstang's small Oxfam Group, led by Bruce Crowther.

Early Campaigning

The Oxfam Group started campaigning on Fairtrade back in 1992, before the Fairtrade mark was even launched. It was not initially an easy task. In 1997 the group tried unsuccessfully to get Fairtrade products used in local cafés and restaurants. They also targeted churches, by giving a large catering tub of Fairtrade instant coffee to each of the six places of worship in Garstang and inviting them to order more. Three took up the offer. Later in 1997 a *Garstang Fairtrade Guide* was published, although at that time only five places in Garstang sold any Fairtrade products.

Making the Local Link

To try to persuade the remaining churches, schools, and traders to use and/or sell Fairtrade products, the group decided to organize a Fairtrade meal at a local restaurant during Fairtrade Fortnight 2000. However, they realized the local campaigning climate had changed. As Bruce Crowther explained, 'when I saw dairy farmers marching down Garstang High Street carrying a banner bearing the words 'We want a fair share of the bottle', I realized that we could no longer continue campaigning on fair trade with developing countries, without the link to local farmers. They also want a fair price for their produce.' To highlight the relevance of fair trade to farmers around rural Garstang, the meal consisted of both local produce and Fairtrade products. The Mayor, head teachers, clergy, traders, and farmers' representatives were invited to the meal. The Mayor became interested in the Fairtrade campaign, the aim of which was now to get Garstang declared a Fairtrade Town, the first in the world.

Place, Power, and Provenance in the Conventional Food System

The conventional food system embraces some of the biggest names in the corporate universe—such as Monsanto and DuPont in food technology, Cargill and ConAgra in food processing, Nestlé and Unilever in food manufacturing, Wal-Mart and Carrefour in food retailing, and McDonald's and

Burger King in food service. Each of these firms constitutes a dense corporate network composed of its own multi-site facilities, multiple suppliers, and strategic partnerships, similar to the 'food clusters' we encountered in Ch. 3. In their different ways, many of these firms see the problems of diet-related disease as offering untapped opportunities for new product development, so much so that health, nutrition, and well-being are thought to offer the best growth prospects in the foreseeable future. Far from being the preserve of the alternative food chain, the health and well-being market is now being targeted by the very same food and drink companies that contributed to the problem of diet-related disease in the first place. Let us take three examples to illustrate this zeitgeist.

Although the majority of its products remain high in fat, sugar, and salt, PepsiCo was quicker than most of its rivals to recognize the market potential of a healthier product portfolio. The company recently introduced its own labelling system in the US to identify healthier products, using criteria set by an independent board of health experts. As a result, some 40 per cent of sales come from products designated with the green 'Smart Spot' given to healthier brands such as sugar-free cola, for example. While PepsiCo's new strategy is largely designed to pre-empt tougher regulation and anti-obesity litigation, the company insists that the new direction is good for business because, in the words of its chairman, Steve Reinemund, 'Smart Spot products grew at more than twice the rate of those without the designation last year, providing more than half the company's revenue growth' (J. Grant and Ward, 2005). To improve its health profile the company acquired two companies with strong 'nutritional images', namely Quaker Oats and Tropicana, which helped to offset the unhealthy image of Pepsi-Cola and Frito-Lay. At the same time, even Frito-Lay is seeking to change its image because, according to its new chief executive, 'one of the great untapped opportunities is to take a leadership role in health and wellness' (ibid).

Nestlé, the largest food company in the world, is thinking along similar lines. Foods with alleged medical benefits, otherwise known as 'functional foods' and 'nutraceuticals', are expected to be the biggest source of growth for the next twenty years. Peter Brabeck, its chief executive, confirmed this when he said that 'it is my conviction that the next value creation, and it will be huge, is going to be nutritional aspects. That is what allows you to ask 40 per cent more for a product' (Benady, 2005).

As a symbol of US-style globalization, and therefore an iconic target for anti-capitalist protesters around the world, McDonald's is perhaps the most controversial conversion to healthy product lines. Ever since its first corporate social responsibility report appeared in 2002, where it committed itself to minimizing ecological damage and improving animal welfare, McDonald's has sought to improve the nutritional image of its products. A combination of anti-obesity litigation, poor financial results, and the advent of more health-conscious consumers forced the company to rethink its strategy. In the biggest overhaul of its product lines in fifty years, the world's largest fast-food chain introduced a range of healthier products in Europe in 2004 as part of a new menu which was described as being 'more Mediterranean than Midwestern'. While launching the new menu, the head of its European operation, Denis Hennequin, bluntly declared that 'McDonald's must move beyond simply providing "convenience" and start catering to a real shift in awareness of the need for a well-balanced diet' (Johnson, 2004). Some European managers at McDonald's had been trying to reform the menu for years, to reduce salt levels, for example, but the company's headquarters in Illinois blocked them, fearing it would change the flavour of the company's core products. One frustrated UK manager complained that 'the Americans don't understand the pressure we are under from the health campaigners' (Revill, 2004).

No part of the food chain is more closely attuned to these market trends than the supermarkets, arguably the most powerful actors in the agri-food system today. As we saw in Ch. 3, the globalization of the supermarkets is a relatively new phenomenon. Belatedly, some supermarkets are 'going global' in two ways at once: first, their supply chains are becoming more globalized, as they seek to sidestep the seasons to offer year-round produce; second, to overcome sluggish markets at home they are seeking to exploit new growth markets abroad, particularly in the developing countries (Busch, 2004; Reardon et al., 2003). The world's top ten food retailers are shown in Table 7.1, which reveals that Wal-Mart, the number one in terms of turnover, comes way down the league table in terms of foreign sales, suggesting that it is still something of a novice in the globalization stakes.

It may be a latecomer on the global scene, but Wal-Mart is rapidly making up for lost time with an aggressive expansion programme around the world, especially in Latin America, Europe, Korea, and China. Some analysts predict that this is part of a new trend in which food retailing will polarize between global and local companies because:

three forces are pushing the top retailers to further globalization: first, the growing sophistication of consumers; second, capital intensification to extract ever-tighter financial returns; and third, the need to get the best price from suppliers in order to stay competitive, *globally sourcing while appearing local*.

(Lang and Heasman, 2004: 164, emphasis added)

The spatial effects of these new food retailing strategies will vary from one supermarket to another, depending on the particular mix of global/local sourcing they adopt in each national market. It will also depend on the kind of *qualities* the supermarket wants to embody, what significance it attaches to the *provenance* of its products, and what type of *consumers* it wishes to attract. This serves to reinforce our argument in Ch. 3, where we challenged the caricatured view of 'supermarkets' as homogenous entities selling standardized products. To illustrate this point in more detail it is

Table 7.1. The top ten food retailers in 2002

Rank	Company	Country	Turnover ($ m.)	No. of countries	Foreign sales (%)
1	Wal-Mart	US	180,787	10	17
2	Carrefour	Fr.	59,690	26	48
3	Kroger	US	49,000	1	0
4	Metro	Ger.	42,733	22	42
5	Ahold	Neth.	41,251	23	83
6	Albertson's	US	36,762	1	0
7	Rewe	Ger.	34,685	10	19
8	Ito Yok (incl. Seven Eleven)	Jap.	32,713	19	33
9	Safeway	US	31,977	3	11
10	Tesco	UK	31,812	9	13

Source: IGD (2002*b*).

worth examining two supermarket strategies, namely Wal-Mart and Waitrose, to show that place, provenance, and power can be mobilized in very different ways within the conventional sector.

From Price to Provenance: A Tale of Two Supermarkets

With 1.4 million employees, Wal-Mart's workforce is now larger than those of General Motors, Ford, General Electric, and IBM combined and, in revenue terms, it is the world's largest company. Its economies of scale are so vast and its macroeconomic effects so extensive, that Wal-Mart's social significance is compared to that of the Ford Motor Co. a hundred years ago. Since opening its first store in Rogers, Arkansas, in 1962, Wal-Mart has shown considerable dexterity in defining its core customers and catering to their needs, for example by providing products that appeal to low-income women. In retrospect, one of the best decisions of Wal-Mart's founder, Sam Walton, was to locate many of its earliest stores in towns with populations of fewer than 5,000 people, communities that were ignored by its rivals. This strategy gave Wal-Mart a near monopoly in its local markets and helped it to weather the recessions of the 1970s and 1980s more successfully than the likes of K-Mart and Sears, its larger competitors at that time (Head, 2004).

Wal-Mart's competitive strengths fall into two categories: its *technological* repertoire and its *social* repertoire. The technological repertoire is believed to have been one of the biggest drivers of US productivity growth in the second half of the 1990s. A study by the McKinsey Global Institute found that the organizational innovations pioneered by Wal-Mart were the key to the step change in productivity in general retailing between 1995 and 2000, and these included 'more extensive use of cross-docking (taking goods directly from factory to store) and better flow of goods/palleting; the use of better forecasting tools to better align staffing levels with demand; redefining store responsibilities and cross-training employees; improvements in productivity measurements and utilisation rates at check-out'. The McKinsey study concluded that 'Wal-Mart demonstrates the impact that managerial innovation and effective use of IT by individual firms can have on market structure, conduct and performance' (London, 2004).

On its own admission, the company's biggest technological advantage is its supply chain, Wal-Mart's core competence. Difficulties in supplying its first rural stores underlined the need for a robust supply chain early in its history, and ever since Wal-Mart has been a leader in supply-chain technologies such as IT, barcode scanning, and, most recently, radio frequency identification (RFID) tags. By 1988, it had the largest privately owned satellite communications network in the US and its central database is second in size only to that of the Pentagon. These technologies enabled Wal-Mart to pioneer data sharing with suppliers through its Retail Link system, which constitutes the most sophisticated collaborative planning, forecasting, and store replenishment system that any retailer operates anywhere in the world. The company uses this information to gain unrivalled insight into consumers' behaviour, which it then shares with suppliers. These technologies helped Wal-Mart to reverse the traditional balance of power in the US between supplier and retailer to the clear advantage of the latter. Even leading suppliers of branded goods have seen the benefits of accessing the largest distribution channel ever created, selling up to 30 per cent of their total volume through Wal-Mart. The buyer–supplier relationship at Wal-Mart is something of a Faustian bargain because, as the volume of a supplier's sales increases, its margins tend to fall, leaving suppliers to live with the trade-off for the sake of volume growth. Overall, however, the general formula for supplying Wal-Mart appears to be disconcertingly simple: according to one supplier, the supermarket 'demands the best possible goods at the lowest price—and nothing else' (Buckley, 2004*b*).

While Wal-Mart's technological repertoire tends to be admired, especially in management circles, its *social* repertoire is often deplored. Among other things, its critics allege that when Wal-Mart arrives in a locality it forces other retailers out of business; it replaces the jobs lost with fewer, lower-paid jobs with longer hours and fewer benefits; it creates more traffic; and it destroys the local character of the community. At the heart of the social critique of Wal-Mart is the fact that the company's celebrated policy of 'Every Day Low Prices' is partly based on a strategy of 'Every Day Low Wages and Benefits'. The average pay of a sales assistant at Wal-Mart in 2003 was about $8.50 an hour, equivalent to $14,000 a year, which was $1,000 below the government's definition of the poverty level for a family of three. As a result, fewer than half its employees can afford even the least-expensive healthcare package offered by the company. According to a recent report by the Democratic Staff of the House Education and Workforce Committee, a 200-employee Wal-Mart store costs the US federal government

$108,000 a year for children's healthcare; $125,000 a year in tax credits and deductions for low-income families; and $42,000 a year in housing assistance. Overall, the cost to the federal taxpayer was put at $420,000 a year for a 200-employee store, equivalent to a total annual welfare bill of $2.5 billion for Wal-Mart's US employees (Buckley, 2004*a*; Head, 2004).

In other words, this social repertoire transfers the social welfare burden from the company on to the state. Low prices may benefit its consumers, but this is partly financed by pay and benefits levels that are significantly below those of rival supermarkets. While Wal-Mart's sales assistants earned less than $9 an hour in 2003, a comparable job at Safeway or Albertson could earn $13 an hour with full healthcare benefits. The hostile reception to Wal-Mart's expansion plans in urban America is largely fuelled by fears that this social repertoire will become the norm for all retailers. This is exactly what provoked an unsuccessful five-month strike among super-market workers in California which followed Wal-Mart's decision to open forty supercentres in the state. The threat of low-wage competition from a company that is famously anti-union caused Safeway to demand pay and benefit cuts from its own staff even before Wal-Mart had opened a single store, a move that forced the Union of Food and Commercial Workers (UFCW) into strike action. Such conflicts are likely to become more com-monplace in the US as Wal-Mart seeks to penetrate the big city grocery markets, all of which are heavily unionized. Although Wal-Mart accepted unions when it acquired Asda in the UK, it is otherwise a union-free com-pany. The only known case of union success in the US occurred in 2000 in a meat-cutting department in a store in Texas, but the employees were imme-diately fired and the department was closed down—both illegal acts under the National Labour Relations Board. These incendiary labour practices will deservedly garner much greater publicity in the largest civil rights case of its kind in US history: the *Dukes* case, a class-action lawsuit on behalf of 1.6 million female employees, past and present, which alleges systematic discrimination against women at Wal-Mart (Featherstone, 2004; Green-house, 2003; Head, 2004).

The global ambitions of Wal-Mart—the 'Beast of Bentonville' to its critics—are of significance to the whole world of food retailing because, given its scale, Wal-Mart can influence the retail environment wherever it operates. Although it entered the food market relatively late in its career, indeed as recently as 1988, it was already the biggest food retailer in the US by 2001, with sales of $53 billion. Even though it currently sources most of its food from within the US, spending some $40 billion a year with domestic agri-food suppliers, this will change as it pursues more global sourcing to sustain its 'everyday' low-price policy. Awesome purchasing power, un-rivalled supply-chain efficiencies, and low pay are the three 'secrets' of Wal-Mart's ability to offer food at prices that are, on average, 15 per cent cheaper than those of its US rivals. Paradoxically, even though its influence as a grocer is second to none, food itself plays a highly instrumental role in its retailing strategy. As one leading US analyst put it: 'Wal-Mart's low food prices entice customers to visit its stores more frequently; and when they do, they're also likely to buy high-margin general merchandise' (Buckley, 2003). Low-cost food, in other words, is a loss-leader to promote more profitable non-food products.

One of the biggest issues at stake in the Wal-Mart story is whether its social repertoire will be sustainable in the US, and whether it is exportable to other countries, especially in Europe. The UFCW has forged a new coalition in the US, the Center for Community and Corporate Ethics, composed of fifty local and national groups, the aim of which is to contest Wal-Mart as it expands into the pro-union urban areas of the US. This civic campaign aims to unite labour and community opposition to a company associated with a 'race to the bottom' because it threatens existing terms and conditions and external-izes the costs of its success by transferring them to the state. Other US states may follow the example of Maryland, where the state assembly is debating a bill that would require Wal-Mart to increase significantly the amount of healthcare coverage it provides to its employees as the retailer is blamed for burdening state Medicaid insurance systems with its uninsured and low-paid workers (Birchall, 2005).

At the other end of the supermarket spectrum stands Waitrose, the leading quality food retailer in the UK. Created in 1904, Waitrose is part of the John Lewis Partnership, the largest and most successful employee-owned business in the country, a corporate form that liberates it from the short-term financial pressures of being a publicly listed company. Although it is the fastest-growing British grocer, Waitrose accounts for just 4 per cent of the UK market, and with only 166 stores and 30,000 employees it is clearly miniscule compared with Wal-Mart. Scale is not the only difference between the two supermarkets; in fact, Waitrose uses a very different corporate metric, as it markets itself on the basis of *provenance*, rather than price. Among its many distinctive features, Waitrose was the first UK supermarket to introduce organic food in 1983 and it won the Organic Supermarket of the Year award on a number of occasions in recognition of its pioneering role in developing the organics market in the UK. It also played a leading role in promoting more sustainable forms of conventional agriculture by sourcing all non-organic crops from Integrated Crop Management Systems, which minimizes the use of chemicals and provides an independently audited system for all fruit and vegetables produced in the UK.

These efforts were rewarded in 2000, when Friends of the Earth ranked Waitrose first in its poll covering GM-free sourcing, the use and restriction of pesticides, and organic commitment. Significantly, its 2004 corporate social responsibility report (Waitrose, 2004) began by emphasizing the provenance of its food:

consumers are naturally concerned by publicity surrounding food scares or food-related diseases, whether BSE, foot and mouth or GM crops. Of late, concerns about childhood obesity have dominated the news. All this can erode consumer confidence. Not surprisingly, customers want to know that the food they are eating is safe and healthy. They want to know where it comes from, how it has been produced and what it contains. Waitrose puts food integrity and traceability at the top of its buying requirements, even though ensuring food integrity across 18,000 product lines and 1,500 suppliers is no easy task. Waitrose relies on long-term relationships with its suppliers, its own inspections and a range of farm assurance schemes to ensure food quality.

Waitrose claims to put these sustainable principles into practice in its supply chain and at Leckford Estate, its own 4,000-acre farm in Hampshire. Leckford aims to produce 'the finest food from known sources', which is the basis for selling 'food you can trust' (ibid.). A critical review of UK supermarket trends concluded that 'nowhere is the greening of the supermarkets more conspicuous than at Leckford', one of the attributes that helped Waitrose to win the Compassionate Supermarket of the Year title in 2003 (Purvis, 2004).

After overtaking the domestic supermarkets that were hitherto regarded as the best for food quality—namely Marks & Spencer and Sainsbury's—Waitrose is now rapidly shedding its image as a purely middle-class, south of England business. With town planners queuing up for a Waitrose store to spearhead urban regeneration, the quality supermarket tends to be feted where Wal-Mart is feared. Apart from the value of its imprimatur as a retail brand name, Waitrose offers more tangible benefits to town and country planners in the form of its commitment to small local producers. Launched in 2002, its Locally Produced range of produce stresses 'fine quality foods selected from your region'. The Locally Produced label includes food and drink made by small-scale producers and supplied to shops within a thirty-mile radius of the production site. Local producers are not required to supply all Waitrose stores, nor do they need to grow their businesses any larger than they wish. The Locally Produced initiative covers some 160 different producers, offering 360 product lines in over 100 stores, more than any other UK supermarket (Waitrose, 2004).

A big expansion plan is under way, which could double Waitrose's store count. One of the aims here is to help the company to shed its narrow class image. Steve Esom, the head of Waitrose, says: 'we are not posh. We do not have one type of customer. Love of food is not dictated by earnings and postcodes. It's about an attitude to life' (Finch, 2005). While the social profile of the Waitrose customer base is more varied than the traditional image might suggest, it is nevertheless disingenuous to suggest that income and geography play no part in the social shaping of food choice. One of the biggest challenges facing Waitrose as it seeks to expand throughout the UK is to prove that it can maintain its reputation for food quality as it doubles its store count—a case of maintaining its distinctive *metric* while it scales up its operations.

The real significance of Waitrose lies in the fact that it furnishes hard evidence that sustainability pays dividends, proving that a conventional supermarket can provide good quality, traceable food while also respecting its partners in the food chain—its employees, suppliers, consumers, and nature. Its success has encouraged other UK supermarkets to explore the sustainability theme; Marks & Spencer, for example, is placing a stronger accent on the traceability of its food by emphasizing that it comes 'from farms we know and trust' (Maitland, 2003). Although the growing resonance of place and provenance in the conventional food system owes much to the success of Waitrose, as we will discuss in the next section, it owes even more to a regulatory climate that is becoming less tolerant of the placeless foodscape.

Place, Provenance, and the Politics of Food Labelling

As consumers and citizens have become more concerned about the safety and quality of their food, they are naturally paying more attention to food labelling, a field which has been described as the new 'battlefield' of food politics (Lang and Heasman, 2004). The notion of 'consumer sovereignty' would seem to be an oxymoron in this field, considering that many consumers, perhaps even the majority, know little about how their food is produced or how it gets from farm to fork. Some supermarkets have publicly admitted that food labelling has traditionally concealed far more than it has revealed. In a damning indictment of food labelling called *The Lie of the Label*, one supermarket exposed a number of morally dubious practices, such as concealing the real nutritional value of products (Co-op, 2001). These loopholes are gradually being tightened up: the US has introduced mandatory nutrition labelling for most food products and the EU is introducing tougher regulations to ensure that labelling, especially with regard to health, is 'clear, accurate and meaningful' (CEC, 2003).

Given the more processed diets of the US and the UK, the concern about fat, sugar, and salt content is more pronounced in these countries than in Italy, for example. In its most recent consumer attitudes survey, the Food Standards Agency in the UK found that consumers were becoming increasingly concerned about their diet and health, with salt, fat, and sugar in the top five food concern issues. However, it also found that concern about the accuracy of food labelling has risen significantly since its first survey in 2000 (FSA, 2005). The biggest obstacle to reducing salt, fat, and sugar in processed food in Anglo-American diets is the processed food industry itself, which has waged successful campaigns in the past to preserve as much of the status quo as possible (Lawrence, 2004; Nestle, 2002). With the obesity epidemic, however, the political climate may be moving against the processed food lobby.

Apart from nutritional issues, the most confusing aspects of food labelling policy revolve around place and provenance, or, in other words, the spatial and social history of the product. Under existing EU regulations for conventional produce, the particulars of the place (not necessarily the country) of origin or provenance of a food must be shown only if failure to do so might mislead a purchaser as to the true origins of the food. Yet, this still leaves room for ambiguity, if not dishonesty. For example, it permits olive oil pressed in Italy from olives grown in Greece to be labelled as 'made in Italy' and it allows beef reared and slaughtered in Brazil, but processed in Europe, to carry a 'product of the EU' label. To study this problem, a parliamentary inquiry in the UK recently examined the whole question of food information policy and it unearthed a paradox: while consumers were becoming increasingly concerned about 'the ethical issues associated with food production', they had 'no means of independently verifying claims made on food labels, or elsewhere, about food production methods'. This led the parliamentary report (House of Commons, 2005) to the following conclusion:

fundamentally, we consider consumers should receive better information about these ethical issues. . . . We appreciate that the scope of legislating for compulsory provision of such information, on either a UK or EU basis, is limited by the WTO Agreements on Technical Barriers to Trade and on the Application of Sanitary and Phytosanitary Measures. Nevertheless, we consider that food producers, manufacturers and processors should consider ways in which they can provide consumers with further information about these matters. Failure to do so could well be interpreted by consumers as a failure to engage with the ethical implications of the industry's activities.

Country of Origin Labelling (COOL) is becoming a major issue in the UK and the US, where smaller producers in particular are trying to overcome their placeless foodscapes by making the food chain more traceable and transparent. In the UK, where the proliferation of farm assurance schemes had begun to confuse consumers, in 2000 a new coalition of farmers and producers launched the Little Red Tractor label as a more generic logo, designed to be 'a tangible endorsement of provenance'. Whatever its merits, the Little Red Tractor label added to the labelling confusion in the UK because most consumers thought it was an indication of origin. In fact, it referred to a *standard* of production, not a *location* of production, and therefore it could be used on imported produce. In its enthusiasm to promote local food, the National Farmers' Union implied that the new logo was exclusive to British produce, but this is illegal under the rules of the EU Single Market. As a result of this experience, the NFU is campaigning for stronger COOL legislation at the EU level.

The COOL issue is proving to be even more controversial in the US, where it has divided the agri-food chain, with small farmers and producers supporting a mandatory labelling system and large agri-business supporting a voluntary system. A provision in the 2002 Farm Bill required grocery stores to identify certain products—beef, pork, and lamb, fish and shellfish, fruits and vegetables, and peanuts—by country of origin. Originally designed to evolve into a mandatory system in 2004, after a two-year trial period as a voluntary scheme, the COOL provision was undermined by the large agri-business lobby, which succeeded in getting the House Agriculture Committee to repeal the mandatory legislation and replace it with a voluntary labelling programme. Since family farms groups had been the major sponsors of a mandatory system, this was a major defeat for the small producer and a major victory for large packers and retailers, many of whom do not believe that such labelling is even necessary. Patrick Boyle, the president of the American Meat Institute, which represents US meat processors, argued that COOL information is not important to consumers. US consumers, he argued, were only interested in information like 'price, freshness and quality', not 'the family tree' issues of place and provenance (Boyle, 2004). US agri-business clearly remains deeply attached to the placeless foodscape. Its opposition to mandatory COOL reflects the fact that it wants the flexibility to source from anywhere in the world without being encumbered by any duty to specify place and provenance.

This 'placeless foodscape' perspective also lay at the root of the US-led campaign in the WTO to overturn the EU system of protection of Geographical Indications (GIs), the labelling scheme that seeks to establish the strongest possible association between the place and provenance of a product (Ilbery and Kneafsey, 2000). Along with Australia, the US had charged that 'GIs kill trademarks' and, therefore, they should be ruled illegal. On 15 March 2005, however, a WTO panel rejected the complaint and ruled that the EU system for protecting these names was compatible with WTO rules and it confirmed that GIs could coexist with prior trademarks (European Commission, 2005). Unlike trademarks, which are privately owned intellectual property rights that can be bought and sold, GIs are a spatially specific public good, in the sense that they protect the geographical name of a product from a given region. The collective legal status of GIs helps small producers, often from poor regions, to gain a form of protection and promotion for their products that is embedded in, and tied to, their region, an asset that, unlike conventional trademarks, cannot be delocalized. GIs are also different from trademarks in that they aim to secure a fairer distribution of added value in supply chains (Barham, 2003; Sylvander et al., 2004).

By protecting the collective rights of the producers of GI-labelled products, this WTO ruling provides a powerful illustration of the interdependence of the local and the global. Perhaps more to the point, it also highlights how and why the local (in this case a local label) is a fragile and unsustainable construct unless it can be defended and promoted at both national and supra-national political levels. As we shall see in the following section, although 'little victories' can and do occur at the local level, these will remain profoundly vulnerable unless they are endorsed and enhanced by wider forms of political support in the multilevel polity.

Place, Power, and Provenance in Alternative Food Networks

Ten years ago it would have been possible to speak of 'alternative food networks' without adding so many caveats because the conventional/alternative distinction seemed so much clearer than it does today. To explore the meanings and implications of alternative food networks (a term widely used in the literature to signify the heterogeneity of the sector), this final section addresses three key themes. First, we provide a brief overview of the alternative food sector to pose a simple, but fundamental, question: in what sense *alternative*? Second, we offer a sympathetic critique of some of the assumptions underlying the *localization* thesis, particularly the claims about social justice and spatial embeddedness. Finally, drawing on the regional vignettes in Chs. 4–6, we offer some speculative proposals as to how a more sustainable agri-food system might be fashioned through new regional ecologies.

In What Sense Are Alternative Food Networks Alternative?

Most alternative food networks (AFNs) seem to have originated as a reaction to some negative trend in the conventional food system, such as the use of pesticides, or in defence of something that was under threat, such as the loss of community food stores. In many cases AFNs have developed in conjunction with the 'quality turn' in food consumption or as part of the renaissance of 'local' food networks (Murdoch, Marsden, and Banks, 2000). From organics and fair trade to regional and artisanal products, what all these AFNs share in common is that they are all examples of 'food markets that redistribute value through the network against the logic of bulk commodity production; that reconvene "trust" between food producers and consumers; and that articulate new forms of political association and market governance' (Whatmore, Stassart, and Renting, 2003).

Although this is very useful as a generic definition, it nevertheless implies that there is a clear boundary between alternative and conventional sectors, when in fact there is not. Perhaps the most dramatic example of the 'conventionalization' of the 'alternative' sector comes from the recent history of organics in California. Generally thought to be the most developed form of sustainable agriculture, organic agriculture is assuming forms that make a mockery of its founding principles. The sober research of Guthman shows quite convincingly that the colonization of organics by agri-business in California is doing much more harm than simply creating a softer path of sustainability, a kind of 'organic lite' as she puts it. Much more important is the fact that 'the conditions it sets undermine the ability of even the most committed producers to practice a purely alternative form or organic farming' (Guthman, 2004). As we saw in Ch. 5, California is the premier state in the US for both conventional and organic food production. Having pioneered some of the worst excesses in conventional agri-industrialization—such as 'feed-lot' dairying, the use of synthetic nitrogen to accelerate crop turnover, and the systematic use of labour contractors, for example—California is now pioneering the intensification of organic agriculture, crystallized in the local maxim of 'get big or get out' (ibid.). Guthman has been criticized for highlighting this 'conventionalization' story, but some of these criticisms suggest a tendency to 'shoot the messenger' for delivering an unwelcome message from the West.

The vast majority of AFNs in Europe and North America seem to be committed to fashion food systems that are environmentally sustainable, economically viable, and/or socially just. However, Allen and her colleagues have posed a fundamental question about the alternative nature of AFNs, namely: 'to what degree do they seek to create a new structural configuration—a shifting of plates in the agrifood landscape—and to what degree are their efforts limited to incremental erosion at the edges of the political-economic structures that currently constitute those plates? That is, are they significantly oppositional or primarily alternative?' (Allen et al., 2003a). Drawing on a survey of alternative agrifood initiatives in California, they found that the 'most striking thing about the responses is the extent to which they accept the structures and parameters of the current food system' (ibid.). Significantly, though, these responses were influenced by the relative lack of institutional support from the state authorities in California.

Another fruitful perspective is that of Watts and his colleagues, who have examined what the concept of 'alternativeness' might mean in the context of recent debates in economic geography. Their analysis is predicated on the idea that 'AFNs can be classified as weaker or stronger on the basis of their engagement with, and potential for subordination by, conventional food supply chains operating in a globalizing, neoliberal polity' (D. C. H. Watts, Ilbery, and Maye, 2005). Watts and colleagues posit the intriguing idea that it is the strength of the *network*, rather than the attributes of the *food*, that provides the strongest form of alternativeness to the conventional system. To illustrate this argument, they focus on a number of different dimensions of alternativeness in short food supply chains (SFSCs), which they consider to be among the stronger networks:

- SFSCs may present a *spatial* alternative to conventional supply chains either by reducing the distance that food travels between production and consumption (the classic examples being farmers' markets, farm shops and box schemes) or by bringing food into places that are poorly served by the conventional system (like 'food deserts');

- SFSCs may present a *social* alternative to conventional chains by forging face-to-face contact between producers and consumers, promoting supply-chain trust and community integration;

- SFSCs may offer an *economic* alternative by creating local markets for local produce which allow the primary producer to capture more value relative to conventional supply chains.   (ibid)

Even though these are all plausible propositions, the underlying argument needs to be qualified in two respects. First, the distinction between the process (the networks) and the product (the food) is rather artificial because, if the quality of the food is poor, the network would quickly atrophy. Second, the argument comes close to celebrating alternativeness as an end in itself, when it was always meant to be a means to more substantive ends—namely, the creation of socially just, economically viable, and environmentally sustainable food chains. Further qualifications can be made of AFNs in the context of 'food system localization'.

Deconstructing the 'Local' in Alternative Food Networks

Over the past decade or so there has been an explosive growth in the number and range of local food initiatives in Europe and North America. While each of these initiatives speaks to the concrete specificity of a 'local' concern, a more general influence seems to be at work as well. In the UK, for example, a striking growth of local community initiatives occurred in the decade following the 1992 Earth Summit in Rio, where local communities were enjoined to make their own contributions to sustainable development. The fact that food-related projects were more popular than all other kinds of projects was attributed to 'a succession of food scares coupled with the growing (if unsurprising) realisation that eating well is good for you' (Shell, 2002). All these local food initiatives were assumed to be wholly positive and benign developments, a tacit assumption that elides the issues of power, privilege, and poverty under the felicitous rubric of the 'local'.

In much of the traditional AFN literature we find the same problem, since 'food system localization is often assumed to be a good, progressive and desirable process' (Hinrichs, 2003: 33). Far from being confined to the AFN literature, the tendency to fetishize the 'local' is common throughout the social sciences, where this is often treated uncritically as something that is more authentic and less threatening than its binary opposite, the 'global'. Thanks to the work of Harvey in particular, some of the more self-critical AFN literature is recognizing the danger of fetishizing the 'local' realm (Allen et al., 2003a; D. Harvey, 1996; Hinrichs, 2003). In the field of agrifood studies, Goodman has posed some of the most searching questions about the social and spatial effects of AFNs, especially regarding the alleged contributions of the latter to social justice and rural development. For example, the class composition of producers and consumers in AFNs leads Goodman to claim that 'alternative quality food production seems destined to retain its status as a narrow "class diet" of privileged income groups', with the result that those who are unable to 'secure access to safe, nutritious food are the missing guests at the table set by this model' (Goodman, 2004: 13).

Not all AFNs are equally open to this critique, as we can see in the cases where AFNs are explicitly formed to combat poverty and social exclusion. As anti-poverty strategies, however, AFNs cannot have more than a modest effect, and there is a danger that they are becoming 'the new philanthropy'. In fact, they are obliged to tackle food poverty on a piecemeal basis through short-term funding for local food projects, a poor surrogate for national state-wide action (Dowler and Caraher, 2003). While these AFNs address the social-class issue identified by Goodman, their lowly status and their insecure funding leave them confined to the margins of society.

Taken together, these criticisms make a powerful case for deconstructing 'food system localization' so as to establish a better understanding of its social composition, its political objectives, and its contribution to sustainable development. One of the most stimulating contributions to the politics of localization comes from Hinrichs, who examined the 'social construction of scale' in Iowa, a bastion of conventional agriculture in the US (Hinrichs, 2003). Hinrichs argues that differing political inflections in food system localization begin with the spatial referent for 'local', but they vary in their assumptions about the boundaries between local and non-local. This, she suggests, leads to two very different tendencies: 'defensive localization' and 'diversity-receptive localization'. Defensive localization imposes rigid boundaries around the spatial 'local' and minimizes internal differences in the name of some 'local good'. Stressing the homogeneity of the local, defensive localization can define itself in patriotic opposition to outside forces. Thus, this form of localization can easily become 'elitist and reactionary, appealing to narrow nativist sentiments' (ibid. 37). Diversity-receptive localization, by contrast, 'sees the local embedded within a larger national or world community, recognizing that the content and interests of "local" are relational and open to change' (ibid.).

Both forms of localization can be discerned in the local food movement in Iowa, especially in the 'Iowa-grown banquet meal', where the 'local' is defined as a state-wide phenomenon, as we noted in Ch. 3. Hinrichs argues that there are both positive and negative aspects to this banquet meal, the most exciting trend in Iowa's food system, according to some local notables. On the positive side, she argues that banquet meals help 're-link food to place and symbolize recovery of some small measure of self-reliance'; furthermore, to the extent that they invite people to think about the provenance of their food, 'such meals engender a wider impact' (ibid. 41). The narrow class composition of the banquet underlines the negative side, however, because the meal tends to be 'a perk of the upper-middle, educated classes—the movers and shakers in local politics and the regional economy' (ibid.). However, since the positives clearly outweigh the negatives, Hinrichs concludes (p. 44) that 'food system localization may remake our troubled world in modest and valuable ways. Recognizing the power—and the perilous trap—of the local is a crucial start'.

Critical engagement with the local in the US has its parallels in the UK, especially in the work of Winter, who argues that food localization signals not so much an 'alternative post-global green future' as a form of 'defensive localism'. In an important contribution to the AFN debate, Winter (2003: 31) concludes that:

it is open to question whether we can equate either the turn to quality or the turn to localism as the first steps towards an alternative food economy which will challenge the dominance of globalised networks and systems of provision and herald a more ecologically sound agriculture sector. We are not suggesting that localism is necessarily and always the conservative force that is perhaps implied by the term defensive localism. There is room for much more research here to uncover the motivations of local purchasers and the consequences of their actions.

Enough has been said for us to be wary of the chain of reasoning that implies that food system localization creates socially embedded food networks that are inherently more just and sustainable than anything in the conventional system. The concept of *embeddedness* has assumed something of a totemic status in the AFN literature, where it is often deployed to express the social relations of trust and regard that temper economic transactions between producers and consumers in alternative food systems. In reality, all economies are socially and culturally embedded to varying degrees, and all are subject to the strong *disembedding* forces of money, capital, and technological change—forces that combine to produce the distinctive capitalist process of creative destruction (Cooke and Morgan, 1998; Sayer, 1997). Localized food systems are not immune to the twin processes of embedding/disembedding. Therefore, AFNs should not be exclusively identified with the former, leaving the conventional system in the grip of the latter. Useful as it is, the concept of embeddedness can lead to an *over*-socialized conception of economic behaviour, which is just as misleading as the *under*-socialized conception of neo-classical economic theory to which it was a response. These criticisms are not designed to devalue the significance of embeddedness as a core concept in food system localization; on the contrary, this becomes more robust if it is part of a conceptual analysis which takes *economic* viability more seriously. As Sayer (1997: 22) has said of the 'cultural turn' in the social sciences, when researchers explain the changing fortunes of sectors without reference to the prosaic issues of costs and cash, 'something is seriously wrong'.

Food-system localization has an important role to play in the construction of more sustainable agri-food ecologies, as we will see later, but the main point to establish now is that AFNs are not immunized against the disembedding economic forces that batter the conventional food sector. In this respect, the dichotomy between alternative and conventional food chains will remain a 'battlefield of knowledge, authority and regulation fought around different levels of embeddedness and socio-technical definitions of quality' (Marsden, 2004*b*). While 'quality' is widely recognized to be one of the key sites of the battle, much less attention has been devoted to another, equally important site: namely, metrics. Just as there is no uniform conception of 'quality', so there is no undisputed metric for economic calculation. Although we have emphasized the significance of economic viability, it must be said that the conventional calculus of profit and loss is too narrow to be used as a framework to assess the true costs of the conventional food system, as we shall see in the next section.

Fashioning Agri-food Ecologies: A Regional Perspective

If AFNs are ever to become something more than the marginal ecological spaces they are today, they will have to engage with, and draw support from, the multilevel governance system that regulates the agri-food system. Mobilizing political support for the cause of localized food chains is therefore far more than a *local* matter. To explore this issue in more detail, we propose to revisit our regional case studies before offering more speculative conclusions about the prospects for agri-food ecologies, that is, agri-food chains that have been designed *as if* sustainable development really mattered.

From a traditional state planning perspective, perhaps the most disconcerting aspect of the 'localized quality' model in Tuscany was that it owed as much to happenstance as to conscious design, with nothing about the process that was planned in advance. The region's marginal status under the CAP-sponsored regime of industrial agriculture had the unintended consequence of leaving intact a diverse range of authentic products that came to be re-evaluated by a later generation of consumers. This consumer base is highly distinctive, composed as it is of both locals and tourists, which makes the local in Tuscany a highly cosmopolitan entity, one that is not available to poorer regions which might otherwise be tempted to 'copy' the Tuscan model. Besides stressing the essentially unplanned *origins* of the localized quality process, we also need to emphasize the key role that regional institutions played in orchestrating and managing the process at a later point. No one should imagine, for example, that the Tuscan model is the product of unbridled market forces. On the contrary, the bottom-up process of localized quality formation was contingent on the existence of state institutions at the regional and local levels. One of the most conspicuous ways in which state institutions promoted local products was through their public procurement policies, with the result that Tuscany has one of the highest levels of *organic* school meals in Italy (Miele et al., 2005).

The Rural Development Plan was therefore the product of this process, rather than the architect of it, and the process through which it was produced, based on intense consultation among all stakeholders, was deemed to be as important as the plan itself because it created a collective sense of ownership. Summarizing the main features of the localized quality process, we would be inclined to stress: (1) a holistic territorial approach, rather than a segmented sectoral approach to agri-food planning; (2) the high premium attached to the principle of subsidiarity in decision-making; and (3) extensive consultation procedures to enhance a common sense of ownership of the regional strategy. As each of these features is internal to the region, it would be wrong to leave the impression that the localized quality process was wholly a regional matter. Less visible, but no less important, is the role of the EU in sustaining the regional model: for example, the concept of multi-functionality at the heart of the Tuscan strategy is part of the reformed CAP, as we saw in Ch. 2. Equally important, it was EU-level action that successfully managed to defend the GI labelling scheme, on which so much of the Tuscan model depends for the certification of its quality products in the WTO. And EU-level support may be needed if Tuscany is successfully to repel the threats posed by the spectre of 'Californiaization'.

For a state that has been described as 'the world's most advanced agricultural zone', a zone that produces a third of the food that Americans eat, California is clearly more than a 'region' in the traditional sense of the term (Walker, 2004). What is perhaps most significant about California is that it is the leading American state in both conventional and alternative food systems. Although the state has been a willing handmaiden for the conventional sector, as we saw in Ch. 5, by comparison the alternative agri-food sector has been starved of both political and financial support. The history of sustainable agriculture and community food security programmes throughout the US is essentially a story of social movements campaigning to get these alternatives endorsed by a reluctant and conservative political establishment, especially in the shape of the USDA. In California, however, there seems to be a growing tendency for alternative agri-food initiatives to receive support from local government. Local scholars have even begun to suggest that 'California provides fertile ground for the development of a progressive alternative agri-food movement' (Allen, 2004: 15). In sharp contrast to Tuscany, where state institutions played a proactive role in shaping the localized food system, private capital has been the driving force in the conventional sector, while social movements fuelled the alternative sector. If the 'policy environment' is an important factor in shaping the development trajectory of a system (Guthman, 2004), then the alternative food sector will first and foremost have to change the political priorities of the Golden State.

A new set of political priorities was one of the main reasons why a new development trajectory occurred in Wales, where the agri-food sector is trying to escape from the ghetto of commodity production. The formation of a new Agri-Food Strategy was largely due to the advent of the National Assembly for Wales, a directly elected government that replaced a non-elected administration that showed little appetite for reform. The process of framing a new Agri-Food Strategy had parallels with Tuscany in the sense that it evolved out of intense debate among the principal stakeholders in the sector. For this reason, it was the first Agri-Food Strategy that enjoyed a sense of collective ownership. Another similarity between the two regions is that Wales has also put sustainable agriculture at the heart of its Agri-Food Strategy, especially in its commitments to organics and GM-free crops. This common vision brought Wales and Tuscany together as founding members of the European Regions Network on Coexistence in Agriculture, a vehicle for those EU regions committed to fighting GM in the name of sustainable agriculture.

For all the similarities, there are, however, at least three major differences between Wales and Tuscany. Most important of all is the customer base for quality products, which is weak in Wales because it is a much poorer region. Second, food cultures are very different, with Wales sharing the UK's predilection for processed and standardized products. Although the Welsh Agri-Food Strategy is seeking to improve the food consumption culture, in public canteens as well as private restaurants, cultural change is inevitably a long-term process. Lastly, the institutions of the regional state are relatively new in Wales, and there is much less experience of working in a territorial as opposed to a sectoral planning framework. Developing a territorial approach will be one of the big challenges for the National Assembly because, as we saw in Ch. 6, its founding constitution contains a legal duty to promote sustainable development through *all* its policies.

In more general terms, new regional agri-food strategies in Europe are one of the consequences of the proliferation of regional governance systems. The rise of the regional realm poses some difficult questions about the nature of this scale of governance. Does devolution of power to the regional scale signal a progressive or a regressive political step? Does the growth of regional governance foster more participative forms of politics and more transparent forms of policy-making or is this just popular rhetoric to conceal the colonization of a new realm by old elites? Do regional governance systems allow regions to design policies more attuned to their own circumstances or do they devolve portfolios rather than power, allowing central governments to divest themselves of responsibility for regional affairs? Finally, does regional mobilization spell a laudable struggle for cultural identity or should it be read as a belated and atavistic response to the levelling imprimatur of globalization? The nature of regionalism, like localism, is always open to question, not least because it is a contested and contingent process. However, the very fact that we continue to debate the meaning of 'regions' and 'regionalism' suggests that, far from being primordial political attachments doomed to be dissolved in the gastric juices of globalization, sub-national territorial allegiances show no sign of withering away (Morgan, 2004*b*).

Just as there are different kinds of localism, so there are also different types of regionalism. Perhaps the best way to assess whether regionalism is progressive or regressive is by looking at its capacity to create or enhance the things we construe to be intrinsically significant—like deeper democratic structures, social and spatial solidarity, the integrity of the public realm, or sustainable development, for example. Although the sub-national regional realm is assuming more significance in Europe, it is merely the 'inside' of a wider process of rescaling the state, the 'outside' being the growth of supra-national scales of governance such as the European Union, NAFTA, and the WTO. Far from being a purely administrative matter, the spatial scale of governance—local, regional, national, or supra-national—is both a medium for, and a product of, political struggle (Swyngedouw, 2000).

In the past, alternative agri-food strategies were overwhelmingly addressed to just one of these scales: the local. If, as Harvey suggests, we take a larger ecological view of our being we have to 'recognize the hierarchical organization of places', and this means acknowledging the limits of purely local action (D. Harvey, 1996: 353): 'the contemporary emphasis on the local, while it enhances certain kinds of sensitivities, totally erases others and thereby truncates rather than emancipates the field of political engagement and action. While we all may have some "place" (or "places") in the order of things, we can never ever be purely "local" beings, no matter how hard we try.' While Harvey's work provides a stunningly concise critique of the limits of the local, it is much less clear as to how one *practically* combines local and non-local action in the multilevel systems that govern us today. The prosaic truth of the matter is that no single spatial scale is alone sufficient if the aim is to fashion something as complex and radical as a sustainable agri-food system. However, supposing that sustainable development really mattered, that it was actually taken seriously in other words, what changes would we need to see at global, national, and sub-national scales?

At the *global* scale, it would mean that the 'non-trade concerns' of developing and developed countries were properly addressed in a new Agreement on Agriculture in the WTO. Developing countries would need at least two concessions: an end to ruinous export dumping, particularly from the EU and the US, and robust SDT provisions to secure their food security. More controversially, perhaps, the EU would want its own 'non-trade concerns' met to allow it to sponsor its multifunctional model of agriculture and rural development—a model that consciously sustains the links among product, place, and provenance.

At the *national* scale, there continues to be much more room for manœuvre than many OECD governments seem to think. Indeed, there is nothing more disempowering than the notion that nation-state governments are powerless in the face of a supposedly ineluctable process of globalization. Three particular national initiatives could do much to fashion a more enabling environment for sustainable agri-food systems. First, a more comprehensive set of *metrics* to assess the external costs of agri-food development would be a long overdue initiative. The external costs of agricultural production in the US in the areas of natural resources, wildlife and ecosystem biodiversity, and human health are in excess of $16 billion a year (Tegtmeier and Duffy, 2004). A different study of the external costs of farming and food miles of a weekly food basket in the UK put the overall figure at more than £6 billion a year (Pretty et al., 2005). External costs of this magnitude raise fundamental questions about the cheap food claims of the conventional food system.

Armed with a more ecologically informed set of metrics, a second important initiative would be to promote food-chain localization through more creative (i.e. sustainable) *public procurement* policies. While, for example, Italy already does this as part of its cultural approach to school meals provision, in placeless foodscapes such as the UK and the US, creative public procurement could be the most important single factor in fashioning food localization. Although the US federal government has backtracked on country of origin labelling generally, the school meals programme, which provides meals to some 28 million children every day, is one of eight federal programmes that require all food purchases to be of domestic origin (GAO, 2003).

Along with metrics (to view and value things differently) and procurement (to fashion markets for locally produced, nutritious food), a third important initiative would be to create more equitable and transparent *supply chains* by providing primary producers with a fairer deal *vis-à-vis* supermarkets, in particular. In the UK, farmers are mired in a climate of fear in their dealings with the multiple retailers, while family farmers in the US are set to become an endangered species. In both cases, there is a desperate need to help primary producers secure a fairer share of the retail value of their products because retailer-led supply chains are anything but free and fair market transactions (Busch, 2004; Marsden, 2004*b*).

Taken together, these national and supra-national initiatives would help to create a more powerful enabling environment in which regional and local initiatives at the *sub-national* scale would have a better chance to flourish. In the context of a more supportive multilevel polity, the regional realm is an appropriate spatial scale at which to promote food-chain localization on ecological grounds (such as lower food miles and fresh produce) and on developmental grounds (such as local knowledge and the cultural commitment to local and regional products). There is a fundamentalist school of thought that maintains that it is a forlorn hope to expect regions from placeless foodscape countries, such as the UK and the US, to develop the localized and embedded agri-food cultures associated with Italy and France, for example.

In reality, despite a hostile national policy environment (until very recently at least), the UK, for example, has witnessed a renaissance of local and regional foodstuffs from both the conventional and the alternative food sectors, confirming the increasingly hybrid nature of the food system. The recent creation of Regional Development Agencies in England will reinforce this process, since these have seized on regional products as a means of reconnecting the region with its heritage and its nature. Similar trends can be discerned in the US, where the forbidding agri-food stereotype conceals phenomenal diversity at local and regional scales. As we have seen, local food is being rediscovered in Iowa, one of the bastions of the placeless foodscape. Even more significant, perhaps, is the Missouri Regional Cuisine Project, which is nurturing 'speciality products based on inputs from particular eco-regions, the retrieval of dishes and foodways that have been lost or obscured, and the celebration of cuisines tightly linked to the givens of a particular environmental "place" and the know-how of its inhabitants' (Barham, Lind, and Jett, forthcoming).

Modest as they are, these examples underline the simple but crucially important point that food cultures are not set in aspic. What fundamentalists forget is that even in Italy, one of the exemplars of a localized food culture, enormous efforts are invested in fashioning these cultural values; in other words, cultural values are socially acquired, rather than genetically inherited. It is no accident that Italy has designed some of the most imaginative food education programmes in Europe, such as *Cultura che Nutre* (Culture that Feeds) for example, which disseminates the principles of a healthy diet in which food is 'symbol, culture, history, respect for the environment and knowledge of the Italian agri-food patrimony and territory' (Morgan and Sonnino, 2005). Such programmes attest to the strong political commitment to creating 'traditional' food culture anew in each generation, a process that helps to fashion knowledgeable consumers who value their food and care about such things as place and provenance.

Not all regions in Europe or the US will be able or willing to move in this direction of course, so we will continue to have different worlds of food which will co-evolve, rather than obliterate each other, in a zero-sum game. Where the difference in diet and taste is the result of informed choice, it becomes part of our cultural diversity; but where consumers and communities are obliged to accept something less than safe, wholesome, and nutritious food because of low income and poor knowledge, then it becomes part of a system of social and spatial inequality. The act of food choice might be an intensely private matter, but the *social* environment of food choice is a question that must be a matter for the public realm. The very least we should expect is that safe, wholesome, and nutritious food is readily available to all—in public canteens as well as fine restaurants, in conventional supermarkets as well as alternative markets—because food is the quintessential test of our collective capacity to fashion sustainable communities.

REFERENCES

Agra Europe (2003*a*) 'Ministers Reach Historic Deal on CAP Reform at Last', *Agra Europe*, 27 June.

—— (2003*b*) 'A CAP Reform Agreement that—Just About—Delivers', *Agra Europe*, 27 June.

Agri-Food Partnership (2004) *Second Organic Action Plan 2005–2010* (Cardiff: Agri-Food Partnership).

ALDEN, E., BUCK, T., and WILLIAMS, F. (2004) 'WTO Rules Against Europe on Food Names', *Financial Times*, 19 November.

ALLAIRE, G. (2004) 'Quality in Economics: A Cognitive Perspective', in M. Harvey, A. McMeekin, and A. Warde (eds.) *Qualities of Food* (Manchester: Manchester University Press), 61–93.

—— and BOYER, R. (1995) *La Grande Transformation de l'agriculture. Lectures conventionnalistes et régulationnistes* (Paris: INRA).

—— and WOLF, S. A. (2004) 'Cognitive Representations and Institutional Hybridity in Agrofood Innovation', *Science, Technology, and Human Values* 29(4): 431–58.

ALLEN, P. (2004) *Together at the Table: Sustainability and Sustenance in the American Agrifood System* (University Park, Pa.: Penn State University Press).

—— and KOVACH, M. (2000) 'The Capitalist Composition of Organic: The Potential of Markets in Fulfilling the Promise of Organic Agriculture', *Agriculture and Human Values* 17(3): 221–32.

—— and SACHS, C. (1991) 'What Do We Want to Sustain? Developing a Comprehensive Vision of Sustainable Agriculture', Issue paper 2 (Santa Cruz, Calif.: The Center for Agroecology and Sustainable Food Systems, University of California, Santa Cruz).

—— FITZSIMMONS, M., GOODMAN, M., and WARNER, K. (2003*a*) 'Shifting Plates in the Agrifood Landscape: The Tectonics of Alternative Agrifood Initiatives in California', *Journal of Rural Studies* 19(1): 61–75.

—— —— —— —— (2003*b*) 'Alternative Food Initiatives in California: Local Efforts Address Systemic Issue', Research Brief 3 (Santa Cruz, Calif.: The Center for Agroecology and Sustainable Food Systems, University of California, Santa Cruz).

ALTIERI, M. A. (1988) 'Beyond Agroecology: Making Sustainable Agriculture Part of a Political Agenda', *American Journal of Alternative Agriculture* 3(4): 142–3.

AMIN, A. (1994) 'The Potential for Turning Informal Economies into Industrial Districts', *Area* 26(1): 13–24.

ANDERSON, J., FEENSTRA, G., and KING, S. (2002) 'Stanislaus County Food System Project' (Davis, Calif.: University of California Sustainable Agriculture Research and Education Program, University of California, Davis).

ARCE, A., and MARSDEN, T. (1993) 'The Social Construction of International Food: A New Research Agenda', *Economic Geography* 69(3): 291–311.

ART (Alliance for Responsible Trade) (2003) 'Letter to the Ministers of Trade Meeting in Sao Paolo', 24 June (Washington, DC: ART).

BAILEY, L. H. (1915) *The Country-Life Movement in the United States* (New York: Macmillan).

BANKS, J. (2001) *Organic Food Supply Chains in Wales: Impacts and Potential.* Department of City and Regional Planning, Working Paper, Cardiff University.

BARHAM, E. (2003) 'Translating Terroir: The Global Challenge of French AOC Labeling', *Journal of Rural Studies* 19(1): 127–38.

—— LIND, D., and JETT, L. (forthcoming) 'The Missouri Regional Cuisines Project: Connecting to Place in the Restaurant', in P. F. Bartlett (ed.) *Urban Place: Reconnecting with the Natural World* (Cambridge, Mass.: MIT).

BARJOLLE, D., and SYLVANDER, B. (1999) 'Some Factors of Success for Origin Labelled Products in Agri-food Supply Chains in Europe', paper presented at the 67th EAAE Seminar, 'The Socioeconomics of Origin Labelled Products in Agri-food Supply Chains: Spatial, Institutional and Co-ordination Aspects', Le Mans, 28–30 October.

BARLING, D. (2004) 'Food Agencies as an Institutional Response to Policy Failure by the UK and the EU', in M. Harvey, A. McMeekin, and A. Warde (eds.), *Qualities of Food* (Manchester: Manchester University Press), 108–28.

—— and LANG, T. (2003) 'A Reluctant Food Policy? The First Five Years of Food Policy under Labour', *Political Quarterly* 74(1): 8–18.

BEATTIE, A. (2005) 'Oxfam Backs Protection for Farmers in Poor Countries', *Financial Times*, 11 April.

BECK, U. (1992) *Risk Society: Towards a New Modernity* (London, Sage).

—— (2001) 'Ecological Questions in a Framework of Manufactured Uncertainties', in S. Seidman and J. C. Alexander (eds.), *The New Social Theory Reader: Contemporary Debates* (London: Routledge), 267–75.

BELASCO, W. J. (1990) *Appetite for Change: How the Counterculture Took on the Food Industry* (New York: Pantheon).

BELL, D., and VALENTINE, G. (1997) *Consuming Geographies: We Are Where We Eat* (London: Routledge).

BELLETTI, G., and MARESCOTTI, A. (1997) 'The Reorganization of Trade Channels of a Typical Product: The Tuscan Extra-virgin Olive-oil', paper presented at 52nd EAAE Seminar 'Typical and Traditional Products: Rural Effect and Agro-industrial Problems', Parma, 19–21 June.

BENADY, A. (2005) 'Nestle's New Flavour of Strategy', *Financial Times*, 22 February.

BERRY, W. (1981) *The Gift of Good Land: Further Essays Cultural and Agricultural* (New York: North Point).

BIRCHALL, J. (2005) 'Wal-Mart under Fire in Maryland on Healthcare Costs', *Financial Times*, 6 April.

BODEN, F., et al. (2002) 'CAP is Something we can be Proud of', *Financial Times*, 23 September.

BOGE, S. (2001) 'Insidious Distance', in C. Petrini and B. Watson (eds.), *Slow Food: Collected Thoughts on Taste, Tradition, and the Honest Pleasure of Food* (White River, Vt.: Chelsea Green).

BOLTANSKI, L., and THEVENOT, L. (1991) *De la justification. Les Économies de la grandeur* (Paris: Gallimard).

BONANNO, A. (1994) 'The Locus of Polity Action in a Global Setting', in Bonanno et al. (eds.) (1994: 251–64).

BONANNO, A., BUSCH, L., FRIEDLAND, W. H., GOUVEIA, L., and MINGIONE, E. (eds.), (1994) *From Columbus to ConAgra: The Globalization of Agriculture and Food* (Lawrence: University Press of Kansas).

BORREGAARD, N., and HALLE, M. (2001) 'Striking a Balance for Trade and Sustainable Development', *World Summit on Sustainable Development Opinion*, May (Geneva: International Institute for Environment and Development).

BOYD, W., and WATTS, M. (1997) 'Agro-industrial Just-in-Time: The Chicken Industry in Postwar American Capitalism', in Goodman and Watts (eds.) (1997: 192–225).

BOYLE, J. P. (2004) 'Make COOL Meat Labeling Voluntary', guest opinion, *Billings Gazette*, 3 August.

BRUNORI, G. (2005) 'Rural Strategy in Tuscany', unpublished manuscript.

—— and ROSSI, A. (2000) 'Synergy and Coherence through Collective Action: Some Insights from Wine Routes in Tuscany', *Sociologia Ruralis* 40(4): 409–23.

BRUSA, L. M. (2003) 'Rural Development, Culture and Environment in Southern Italy', unpublished Ph.D. thesis, University of Manchester.

BUCK, D., GETZ, C., and GUTHMAN, J. (1997) 'From Farm to Table: The Organic Vegetable Commodity Chain of Northern California', *Sociologia Ruralis* 37(1): 3–20.

BUCKLEY, N. (2003) 'Wal-Mart Finds Strength in a Smaller Scale', *Financial Times*, 9 January.

—— (2004a) 'Wal-Mart Under Fire: Showdown in the Big-box Shopping Aisles', *Financial Times*, 6 July.

—— (2004b) 'Wal-Mart Under Fire: The Price of Huge Sales and Tiny Margins', *Financial Times*, 7 July.

BUSCH, L. (2004) 'The Changing Food System: From Markets to Networks', paper presented at XI World Congress of International Rural Sociological Association, Trondheim, Norway, July.

BUTLER, L. J., and WOLF, C. (2000) 'California Dairy Production: Unique Policies and Natural Advantages', in H. K. Schwarzweller and A. P. Davidson (eds.), *Dairy Industry Restructuring* (Greenwich, Conn.: JAI), 141–61.

BUTTEL, F. H. (1997) 'Some Observations on Agro-food Change and the Future of Agricultural Sustainability Movements', in Goodman and Watts (eds.) (1997: 344–65).

—— (2003) 'Ever Since Hightower: The New Politics of Agricultural Research Activism in the Molecular Age', paper prepared for presentation at the annual meeting of the American Sociological Association, Atlanta, 16 August.

—— (in press) 'Sustaining the Unsustainable: Agrofood Systems and Environment in the Modern World', in P. Cloke, T. Marsden, and P. Mooney (eds.), *Handbook of Rural Studies* (London: Sage).

—— LARSON, O. F., and GILLESPIE Jr., G. W. (1990) *The Sociology of Agriculture*, (New York: Greenwood).

CAFOD (Catholic Agency for Overseas Development) (2003) 'Memorandum Submitted by CAFOD', House of Commons, *Trade and Development at the WTO: Issues for Cancún*, Select Committee on International Development, oral and written evidence, HC 400-II (London: Stationery Office).

CALLON, M. (1991) 'Techno-economic Networks and Irreversibility', in J. Law (ed.), *A Sociology of Monsters: Essays on Power, Technology and Domination* (London: Routledge), 132–61.

—— and LATOUR, B. (1981) 'Unscrewing the Big Leviathan: How Actors Macro-structure Reality and How Sociologists Help Them to Do So', in K. Knorr-Cetina and A. Cicourel (eds.), *Advances in Social Theory: Towards an Integration of Micro- and Macro-sociologies* (London: Routledge & Kegan Paul), 277–303.

CAPATTI, A. (1999) 'The Traces Left by Time', *Slow* 17: 4–6.

CARMAN, H. F., COOK, R., and SEXTON, R. J. (2004) 'Marketing California's Agricultural Production', in J. Siebert (ed.), *California Agriculture: Dimensions and Issues* (Oakland, Calif.: ANR Communication Services), 89–119.

CARTER, C. A., CHALFANT, J. A., and GOODHUE, R. E. (2002) 'Economic Analysis of the January 2001 California Departmen of Pesticide Regulation Regulations on Strawberry Field Fumigation', report prepared for the California Department of Food and Agriculture (Davis, Calif.: Department of Agricultural and Resource Economics, University of California, Davis).

—— CHALFANT, J. A., GOODHUE, R. E., and MCKEE, G. J. (2005) 'Costs of 2001 Methyl Bromide Rules Estimated for California Strawberry Industry', *California Agriculture* 59(1): 41–6.

CAVALCANTI, J. S. B., and MARSDEN, T. K. (2005) 'The Globalised Fruiticulture System: New Structures and Agency in the North/South Relationships', paper presented at Mini-conference on 'Resistance and Agency in Contemporary Agriculture and Food: Empirical Cases and New Theories', Austin, Texas, 13–14 June.

CEC (Commission of the European Communities) (2003) Proposal for a Regulation of the European Parliament and of the Council on Nutrition and Health Claims Made on Foods, Brussels.

CIAA (Confederation of the Food and Drink Industries of the EU) (2003) *Data and Trends of the EU Food and Drink Industry* (Brussels: CIAA).

CIUFFOLETTI, Z. (ed.) (2002) *L'Uomo e la Terra: Paesaggio Agrario e Prodotti di Qualità* (Florence: Fratelli Alinari).

CLARKE, A. (2003) 'It's Official: Chefs Pronounce Welsh Black as Tastiest Beef', *Western Mail*, 22 March.

COCHRANE, W. W. (1993) *The Development of American Agriculture: A Historical Analysis* (2nd edn., Minneapolis: University of Minnesota Press).

COLMAN, D., and HARVEY, D. (2004) *The Future of UK Dairy Farming* (London: Milk Development Council).

Competition Commission (2000) *Supermarkets: A Report on the Supply of Groceries from Multiple Stores in the United Kingdom*, Cm 4842 (London: HMSO).

COOKE, P., and MORGAN, K. (1998) *The Associational Economy: Firms, Regions and Innovation* (Oxford: Oxford University Press).

Co-op (2001) *The Lie of the Label* (Manchester: CWS).

Cooperative Development Board (1987) 'The Failure of Welsh Quality Lambs Ltd and the Lessons for the Future' (Cardiff: Cooperative Development Board).

COTTERILL, R. W. (1997) 'The Food Distribution System of the Future: Convergence Towards the US or UK Model?', *Agribusiness* 13(2): 123–35.

Cozad, S., King, S., Krusekopf, H., Prout, S., and Feenstra, G. (2002) 'Alameda County Foodshed Report' (Davis, Calif.: University of California Sustainable Agriculture Research and Education Program, University of California, Davis).

Cronin, W. (1991) *Nature's Metropolis: Chicago and the Great West* (New York: W. W. Norton & Co.).

de Janvry, A. (1981) *The Agrarian Question and Reformism in Latin America* (Baltimore, Md.: Johns Hopkins University Press).

de Jonquieres, G., et al. (2003) 'Sowing Discord: After Iraq, the US and Europe Head for a Showdown over Genetically Modified Crops', *Financial Times*, 14 May.

de Roest, K. (2000) *The Production of Parmigiano-Reggiano Cheese: The Force of an Artisanal System in an Industrialised World* (Assen: Royal Van Gorcum).

Dicken, P. (1998) *Global Shift: Transforming the World Economy* (3rd edn., London: Paul Chapman).

Dowler, E., and Caraher, M. (2003) 'Local Food Projects: The New Philanthropy?', *Political Quarterly* 74(1): 57–65.

DTZ (2004) *Agri-Food Strategy Review 2004–2007* (Cardiff: DTZ).

Dube, S. (2004) 'Dairy Producers Face "Third World" Prices', *Western Mail*, 18 May.

DuPuis, E. M. (2000) 'Not in My Body: rBGH and the Rise of Organic Milk', *Agriculture and Human Values* 17(3): 285–95.

—— (2002) *Nature's Perfect Food: How Milk Became America's Drink* (New York: New York University Press).

—— and Goodman, D. (2004) 'Should We Go 'Home' to Eat? Toward a Reflexive Politics of Localisms', paper presented at the XI World Congress of the International Rural Sociological Association, Trondheim, Norway, July.

Dunn, J. A. (1995) Organic Food and Fibre: An Analysis of 1994 Certified Production in the United States. (Washington, DC: USDA Agricultural Marketing Service).

Ekins, P. (1997) 'Sustainable Wealth Creation in the Local Economy', in G. Haughton (ed.), *Community Economic Development* (London: Stationery Office), 18–21.

European Commission (1987) 'Perspective for the Common Agricultural Policy' (Brussels).

—— (1991) *The Development and Future of the CAP: Reflections Paper of the Commission, Communication of the Commission to the Council*, COM (1991) 100 final (Luxembourg: Office for Official Publications of the European Communities).

—— (2003) *EU Agriculture and the WTO: Doha Development Agenda, Cancún September 2003* (Brussels: European Commission Directorate-General for Agriculture).

—— (2005) WTO Panel Upholds EU System of Protection of Geographical Indications (Brussels).

Eymard-Duvernay, F. (1989) 'Conventions de qualité et forms de coordination', *Revue Economique* 49(2): 329–59.

Fairtrade Foundation (2005) *Fairtrade Towns Initiative* (London: Fairtrade Foundation).

Featherstone, L. (2004) *Selling Women Short: The Landmark Battle for Workers' Rights at Wal-Mart* (New York: Basic Books).

Fenn, D. (ed.) (2004) *Market Review 2004: The Food Industry* (Hampton: Key Note).

FERNANDEZ-ARMESTO, F. (2001) *Civilizations: Culture, Ambition, and the Transformation of Nature* (New York: Free Press).

Financial Times (2002) 'A shoddy form deal', 28 October.

FINCH, J. (2005) 'Lord of the Aisles', *The Guardian*, 2 April.

FINE, B. (1994) 'Towards a Political Economy of Food', *Review of International Political Economy* 1(3): 519–45.

—— and LEOPOLD, E. (1993) *The World of Consumption* (London: Routledge).

FISCHLER, F. (1999), 'Vision for European Agriculture', International Green Week, Berlin, 8 September.

FITZSIMMONS, M. (1997) 'Regions in Global Context? Restructuring, Industry, and Regional Dynamics', in Goodman and Watts (eds.) (1997: 158–65).

—— and GOODMAN, D. (1998) 'Incorporating Nature: Environmental Narratives and the Reproduction of Food', in B. Braun and N. Castree (eds.), *Remaking Reality: Nature at the Millennium* (London: Routledge), 194–220.

FLANDRIN, J.-L. (1999) 'From Dietetics to Gastronomy: The Liberation of the Gourmet', in J.-L. Flandrin and M. Montanari (eds.), *Food: A Culinary History from Antiquity to the Present* (New York: Columbia University Press), 418–32.

—— MARSDEN, T., and HARRISON, M. (1999) 'The Regulation of Food in Britain in the 1990s', *Policy and Politics* 27(4): 435–46.

—— —— and SMITH, E. (2003) 'Food Regulation and Retailing in a New Institutional Context', *Political Quarterly* 74(1): 38–46.

—— CARSON, L., LEE, B., MARSDEN, T., and THANKAPPAN, S. (2004) 'The Food Standards Agency: Making a Difference?', Working Paper 21 (Cardiff: Centre For Business Relationships, Accountability, Sustainability and Society, Cardiff University).

FRIEDLAND, W. H. (1984) 'Commodity System Analysis: An Approach to the Sociology of Agriculture', in H. Schwartzweller (ed.), *Research in Rural Sociology and Development* (Greenwich, Conn: JAI), 221–35.

—— (1994) 'The New Globalization: The Case of Fresh Produce', in A. Bonanno, L. Busch, W. H. Friedland, L. Gouveia, and E. Mingione (eds.), *From Columbus to ConAgra: The Globalization of Agriculture and Food* (Lawrence: University Press of Kansas), 210–31.

—— (1997) ' "Creating Space for Food" and "Agro-industrial Just-in-time" ', in Goodman and Watts (eds.) (1997: 226–32).

—— (1998) 'Book Review of 'Strawberry Fields: Politics, Class, and Work in California Agriculture' by Miriam J.Wells', *Agriculture and Human Values* 15(2): 185–8.

—— (2001) 'Reprise on Commodity Systems Methodology', *International Journal of Sociology of Agriculture and Food* 9(1): 82–103.

—— (2002) 'Agriculture and Rurality: Beginning the Final Separation?', *Rural Sociology* 67(3): 350–71.

—— BARTON, A. E., and THOMAS, R. J. (1981) *Manufacturing Green Gold: Capital, Labor, and Technology in the Lettuce Industry* (New York: Cambridge University Press).

—— BUSCH, L., BUTTEL, F., and RUDY, A. (eds.) (1991) *Towards a New Political Economy of Agriculture* (Boulder: Westview).

FRIEDMANN, H., and MCMICHAEL, P. (1989) 'Agriculture and the State System: The Rise and Decline of National Agricultures, 1870 to the Present', *Sociologia Ruralis* 29(2): 93–117.

FSA (Food Standards Agency) (2005) *Consumer Attitudes Survey* (London: FSA).

GAO (General Accounting Office) (2003) *Country of Origin Labeling*, GAO-03-780 (Washington, DC: General Accounting Office).

GILBERT, J., and AKOR, R. (1988) 'Increasing Structural Divergence in US Dairying: California and Wisconsin since 1950', *Rural Sociology* 53(1): 56–72.

—— and WEHR, K. (2003) 'Dairy Industrialization in the First Place: Urbanization, Immigration, and Political Economy in Los Angeles County, 1920–1970', *Rural Sociology* 68(4): 467–90.

GILG, A. W., and BATTERSHILL, M. (1998) 'Quality Farm Food in Europe: A Possible Alternative to the Industrialised Food Market and to Current Agri-environmental Policies: Lessons from France', *Food Policy* 23(1): 25–40.

GLIESSMAN, S. R. (ed.) (1990) *Agroecology: Researching the Ecological Basis for Sustainable Agriculture* (New York: Springer Verlag).

GOLDSCHMIDT, W. R. (1947) *As You Sow: Three Studies in the Social Consequences of Agribusiness* (Montclair, NJ: Allanheld, Osmun & Co.).

GOODMAN, D. (1991) 'Some Recent Tendencies in the Industrial Reorganization of the Agri-food System', in Friedland et al. (eds.) (1991: 37–64).

—— (1999) 'Agro-food Studies in the 'Age of Ecology': Nature, Corporeality, Bio-Politics', *Sociologia Ruralis* 39(1): 17–38.

—— (2000) 'The Changing Bio-politics of the Organic: Production, Regulation, Consumption', *Agriculture and Human Values* 17(3): 211–13.

—— (2003) 'The Quality "Turn" and Alternative Food Practices: Reflections and Agenda', *Journal of Rural Studies* 19(1): 1–17.

—— (2004) 'Rural Europe Redux? Reflections on Alternative Agro-food Networks and Paradigm Change', *Sociologia Ruralis* 44(1): 3–16.

—— and REDCLIFT, M. (1991) *Refashioning Nature: Food, Ecology and Culture* (London: Routledge).

—— and WATTS, M. (1994) 'Reconfiguring the Rural or Fording the Divide? Capitalist Restructuring and the Global Agro-food System', *Journal of Peasant Studies* 22(1): 1–49.

—— —— (eds.) (1997) *Globalizing Food: Agrarian Questions in Global Restructuring* (London: Routledge).

—— SORJ, B., and WILKINSON, J. (1987) *From Farming to Biotechnology: A Theory of Agro-industrial Development* (Oxford: Basil Blackwell).

GRANT, A. (2005) 'Food Groups Get a Taste of Fear', *Financial Times*, 24 February.

GRANT, J., and WARD, A. (2005) 'A Better Model? Diversified Pepsi Steals Some of Coke's Sparkle', *Financial Times*, 28 February.

GREEN, D., and GRIFFITH, M. (2002) 'Dumping on the Poor: The CAP, the WTO and International Development' (London: Cafod).

GREENHOUSE, S. (2003) 'Wal-Mart, Driving Workers and Supermarkets Crazy', *The New York Times*, 19 October.

GRIFFITHS, S., and WALLACE, J. (1998) *Consuming Passions: Food in the Age of Anxiety* (London: Mandolin).

GRONOW, J., and WARDE, A. (eds.) (2001) *Ordinary Consumption* (London: Routledge).

GUTHEY, G. T., GWIN, L., and FAIRFAX, S. (2003) 'Creative Preservation in California's Dairy Industry', *The Geographical Review* 93(2): 171–92.

GUTHMAN, J. (2003) 'Fast Food/Organic Food: Reflexive Tastes and the Making of 'Yuppie Chow' ', *Social and Cultural Geography* 4(1): 45–58.

—— (2004) 'The Trouble with "Organic Lite" in California: A Rejoinder to the "Conventionalisation" Debate', *Sociologia Ruralis* 44(3): 301–16.

HANF, C.-H., and HANF, J. H. (2004) 'Internationalization of Food Retail Firms and Its Impact on Food Suppliers', paper presented at the 88th EAAE Seminar on 'Retailing and Producer–Retailer Relationship in the Food Chain', Paris, 5–6 May.

HARVEY, D. (1996) *Justice, Nature and the Geography of Distance* (Oxford: Blackwell).

HARVEY, M., MCMEEKIN, A., and WARDE, A. (eds.) (2004) *Qualities of Food* (Manchester: Manchester University Press).

HAYES, D. J., LENCE, S. H., and STOPPA, A. (2003) 'Farmer-owned Brands?', Briefing Paper 02-BP 39 (Ames, Iowa: Center for Agriculture and Rural Development, Iowa University).

HEAD, S. (2004) 'Inside the Leviathan', *The New York Review of Books*, 16 December.

HEFFERNAN, W., and CONSTANCE, D. (1994) 'Transnational Corporations and the Globalization of the Food System', in Bonanno et al. (eds.) (1994: 48–64).

—— HENDRICKSON, M., and GRONSKI, R. (1999) 'Consolidation in the Food and Agriculture System', http://www.foodcircles.missouri.edu/pub.htm (accessed 25 May 2005).

HENDERSON, G. L. (1998) *California and the Fictions of Capital* (New York: Oxford University Press).

HENDRICKSON, M. K., and HEFFERNAN, W. D. (2002) 'Opening Spaces through Relocalization: Locating Potential Resistance in the Weaknesses of the Global Food System', *Sociologia Ruralis* 42(4): 347–69.

HIGHTOWER, J. (1973) *Hard Times, Hard Tomatoes* (Cambridge, Mass,: Schenkman).

HINES, C. (2000) *Localization: A Global Manifesto* (London: Earthscan).

HINRICHS, C. C. (2003) 'The Practice and Politics of Food System Localization', *Journal of Rural Studies* 19(1): 33–45.

House of Commons (2002) *The Future of UK Agriculture in a Changing World*, Environment, Food and Rural Affairs Committee, Ninth Report of Session 2001–02, HC 550 (London: Stationery Office).

—— (2003a) *Trade and Development at the WTO: Issues for Cancún*, International Development Committee, Seventh Report of Session 2002–03, Volume 1, HC 400-1 (London: Stationery Office).

—— (2003b) *Trade and Development at the WTO: Learning the Lessons of Cancún to Revive a Genuine Development Round*, International Development Committee, First Report of Session 2003–04, Volume 1, HC 92–1 (London: Stationery Office).

—— (2004) *Milk Pricing in the United Kingdom*, Environment, Food and Rural Affairs Committee, Ninth Report of Session 2003–2004, HC 335 (London: Stationery Office).

—— (2005) *Food Information*, Environment, Food and Rural Affairs Committee, Seventh Report of Session 2004–2005, HC 469 (London: Stationery Office).

HUGHES, A. (1996) 'Forging New Cultures of Food Retailer–Manufacturer Relations?', in N. Wrigley and M. Lowe (eds.), *Retailing, Consumption and Capital: Towards a New Retail Geography* (Harlow: Longman), 90–115.

ICTSD (International Centre for Trade and Sustainable Development) (2004) 'Special and Differential Treatment' (Geneva: International Centre for Trade and Sustainable Development).

IGD (2002*a*) *Local and Regional Food* (Watford: Institute of Grocery Distribution).

—— (2002*b*) *Global Retailing* (Watford: Institute of Grocery Distribution).

IKERD, J. (2002) 'New Farm Bill and U.S. Trade Policy: Implications for Family Farms and Rural Communities', paper presented at 'Grain Place' Farm Tour and Seminar, Aurora, NB, 27 July.

ILBERY, B., and KNEAFSEY, M. (2000) 'Registering Regional Speciality Food and Drink Products in the UK: The Case of PDOs and PGIs', *Area* 32(3): 317–25.

INOUYE, J., and WARNER, K. D. (2001) 'Plowing Ahead: Working Social Concerns into the Sustainable Agriculture Movement', White Paper (Santa Cruz, Calif.: California Sustainable Agriculture Working Group).

JELINEK, L. J. (1982) *Harvest Empire: A History of California Agriculture* (2nd edn., San Francisco, Calif: Boyd & Fraser).

JOHNSON, J. (2004) 'Healthier Menus at McDonald's', *Financial Times*, 9 March.

Jones, A. (2001) *Eating Oil: Food in a Changing Climate* (London: Sustain).

JONES, C. (2003) 'Minority Party Debate: Genetically Modified Organisms', 24 October (Cardiff: National Assembly for Wales).

—— (2004) 'Farming Connect is Envy of Europe', *Western Mail*, 27 April.

JONES, D., and PRESCOTT, C. (2005) 'An Investigation to Look at Agriculture in Santa Cruz County', in A. Plater, K. Willis, M. Pelling, and A. Morse (eds.), *Santa Cruz Field Course 2002*, Research Report No. 2 (Liverpool: Department of Geography, University of Liverpool), 90–3.

KAUTSKY, K. (1988) *The Agrarian Question* (London: Zwan). First published 1899.

KLOPPENBURG, J., HENDRICKSON, M., and STEVENSON, G. W. (1996) 'Coming into the Foodshed', *Agriculture and Human Values* 13(3): 33–42.

KPMG (2003) 'Price and Profitability in the British Dairy Chain', Report to the Milk Development Board (London: KPMG).

KREBS, J. (2000) *The Food Standards Agency: A Vision for the Future* (Abbots Langley: The Caroline Walker Trust).

—— (2003) 'Is Organic Food Better For You?', speech given to Cheltenham Science Festival, 4 June.

KROPOTKIN, P. (1906) *The Conquest of Bread* (London: G. P. Putnam's Sons).

LAMONT, M., and THEVENOT, L. (2000) 'Introduction: Toward a Renewed Comparative Cultural Sociology', in M. Lamont and L. Thevenot (eds.), *Rethinking Comparative Cultural Sociology: Repertoires of Evaluation in France and the United States* (Cambridge: Cambridge University Press), 1–22.

LANG, T., and HEASMAN, M. (2004) *Food Wars: The Global Battle for Minds, Mouths and Markets* (London: Earthscan).

—— and RAYNER, G. (2003) 'Food and Health Strategy in the UK: A Policy Impact Analysis', *Political Quarterly* 74(1): 66–75.

LATOUR, B. (1999) *Pandora's Hope: Essays on the Reality of Science Studies* (London: Harvard University Press).

LAW, J. (1994) *Organizing Modernity* (Oxford: Blackwell).

LAW, J., and HASSARD, J. (eds.) (1998) *Actor Network Theory and After* (Oxford: Blackwell).

LAWRENCE, F. (2004) *Not on the Label: What Really Goes into the Food on Your Plate* (London: Penguin).

LEITCH, A. (2003) 'Slow Food and the Politics of Pork Fat: Italian Food and European Identity', *Ethnos* 68(4): 437–62.

LEVENSTEIN, H. (1999) 'The Perils of Abundance: Food, Health, and Morality in America's Food', in J. P. Flandrin and M. Montarari (eds.), *Food: A Cultural History* (New York: Columbia University Press).

LIEBMAN, E. (1983) *California Farmland: A History of Large Agricultural Land Holdings* (Totowa, NJ: Rowman & Allanheld).

LILLISTON, B., and CUMMINS, R. (1998) 'Organic vs. "Organic": The Corruption of a Label', *The Ecologist* 28(4): 195–9.

—— and RITCHIE, N. (2000) 'Freedom to Fail: How US Farming Policies Have Helped Agribusiness and Pushed Family Farmers toward Extinction', *Multinational Monitor* 21(7/8).

LOBAO, L., and MEYER, K. (2001) 'The Great Agricultural Transition: Crisis, Change, and Social Consequences of Twentieth Century US Farming', *Annual Review of Sociology* 27: 103–24.

LONDON, S. (2004) 'Wal-Mart Under Fire: A Country within a Country', *Financial Times*, 9 July.

LOWE, P., BULLER, H., and WARD, N. (2002) 'Setting the Next Agenda? British and French Approaches to the Second Pillar of the CAP', *Journal of Rural Studies* 18(1): 1–17.

—— MARSDEN, T., and WHATMORE, S. (1994) 'Changing Regulatory Orders: The Analysis of the Economic Governance of Agriculture', in P. Lowe, T. Marsden, and S. Whatmore (eds.), *Regulating Agriculture* (London: David Fulton), 1–30.

McMICHAEL, P. (1993) 'World Food System Restructuring under a GATT Regime', *Political Geography* 12(3): 198–214.

—— (ed.) (1994) *The Global Restructuring of Agro-food Systems* (Ithaca, NY: Cornell University Press).

—— (1998) 'Global Food Politics', *Monthly Review* 50(3): 97–122.

MAITLAND, A. (2003) 'M&S Checks out a Healthier Shopping List', *Financial Times*, 4 December.

—— (2005) 'Watchdog Need Not Be a Poodle', *Financial Times*, 10 February.

MANN, S. A., and DICKENSON, J. M. (1978) 'Obstacles to Development of a Capitalist Agriculture', *Journal of Peasant Studies* 5(4): 466–81.

MARSDEN, T. (1988) 'Exploring Political Economy Approaches in Agriculture', *Area*, 20(4): 315–22.

—— (1997) 'Creating Space for Food: The Distinctiveness of Recent Agrarian Development', in Goodman and Watts (eds.) (1997: 169–91).

—— (2003) *The Condition of Rural Sustainability* (Assen: Royal Van Gorcum).

—— (2004*a*) 'The Quest for Ecological Modernisation: Re-spacing Rural Development and Agri-food Studies', *Sociologia Ruralis* 44(2): 129–46.

MARSDEN, T. (2004*b*) 'Theorising Food Quality: Some Key Issues in Understanding Its Competitive Production and Regulation', in Harvey, McMeekin, and Warde (eds.) (2004: 129–55).

—— and ARCE, A. (1995) 'Constructing Quality: Emerging Food Networks in the Rural Transition', *Environment and Planning A* 27(8): 1261–79.

MARSDEN, T. and SMITH, E. (2005) 'Ecological Entrepreneurship: Sustainable Development in Local Communities through Quality Food Production and Local Branding', *Geoforum* 36(4).

—— and SONNINO, R. (2005) 'Rural Development and Agri-food Governance in Europe: Tracing the Development of Alternatives', in V. Higgins and G. Lawrence (eds.), *Agricultural Governance: Globalization and the New Politics of Regulation* (London: Routledge).

—— and WRIGLEY, N. (1995) 'Regulation, Retailing, and Consumption', *Environment and Planning A* 27(12): 1899–912.

—— MUNTON, R. J. C., WARD, N., and WHATMORE, S. (1996) 'Agricultural Geography and the Political Economy Approach: A Review', *Economic Geography* 72(4): 361–75.

—— FLYNN, A., and HARRISON, M. (2000) *Consuming Interests: The Social Provision of Foods* (London: UCL).

MASON, L., and BROWN, C. (1999) *Traditional Foods of Britain: An Inventory* (Totnes: Prospect Books).

METER, K., and ROSALES, J. (2001) *Finding Food in Farm Country—the Economics of Food and Farming in South-East Minnesota* (Minneapolis: University of Minnesota Press).

MIDMORE, P. (2004) 'Radical Change Needed to Give Family Farms a Future', *Western Mail*, 10 February.

MIELE, M., et al. (2005) 'Organic School Meals in Italy' (Cardiff: The Regeneration Institute, Cardiff University).

Milk Development Council (2004) *Dairy Supply Chain Margins 2003–04: Who Made What in the Dairy Industry and How It Has Changed* (London: MDC)

MILLSTONE, E. (2005) 'MAFF record on BSE not so open', *Financial Times*, 17 February.

—— and LANG, T. (eds.) (2003) *The Atlas of Food: Who Eats What, Where and Why* (London: Earthscan).

—— and ZWANENBERG, P. VAN (2001) 'The Politics of Expert Advice: Lessons from the Early History of the BSE Saga', *Science and Public Policy* 28(2): 99–112.

—— —— (2002) 'The Evolution of Food Safety Policy-Making Institutions in the UK, EU and Codex Alimentarius', *Social Policy and Administration* 36(6): 593–609.

MINTZ, S. (1979) 'Time, Sugar and Sweetness', *Marxist Perspectives* 2: 56–73.

MITCHELL, D. (2000) 'The Difference that Space Makes in California Agriculture', *Journal of Historical Geography* 26(3): 469–80.

—— INGCO, M. D., and DUNCAN, R. C. (1997) *The World Food Outlook: Waiting for Malthus* (Cambridge: Cambridge University Press).

MONTANARI, M. (1996) *The Culture of Food* (Oxford: Blackwell).

MORGAN, K. (2002) 'Subsidiarity without Solidarity? The Rural Challenge in Post-Devolution Britain', in J. Adams and P. Robinson (eds.), *Devolution in Practice* (London: IPPR), 140–4.

—— (2004*a*) *School Meals: Healthy Eating and Sustainable Food Chains* (London: The Caroline Walker Trust).

—— (2004*b*) 'Sustainable Regions: Governance, Innovation and Scale', *European Planning Studies* 12(6): 871–89.

—— and MORLEY, A. (2002) 'Re-localising the Food Chain: The Role of Creative Public Procurement' (Cardiff: The Regeneration Institute, Cardiff University).

—— and MUNGHAM, G. (2000) *Redesigning Democracy: The Making of the Welsh Assembly* (Bridgend: Seren).

—— and MURDOCH, J. (2000) 'Organic vs. Conventional Agriculture: Knowledge, Power and Innovation in the Food Chain', *Geoforum* 31(2): 159–73.

—— and SONNINO, R. (2005) 'Catering for Sustainability: The Creative Procurement of School Meals in Italy and the UK' (Cardiff: The Regeneration Institute, Cardiff University).

MURDOCH, J. (1994) 'Some Comments on "Nature" and "Society" in the Political Economy of Food', *Review of International Political Economy* 1(3): 571–6.

—— and MIELE, M. (1999) ' "Back to Nature": Changing Worlds of Production in the Food Sector', *Sociologia Ruralis* 39(4): 465–83.

—— —— (2004) 'A New Aesthetic of Food? Relational Reflexivity in the "Alternative" Food Movement', in Harvey, McMeekin, and Warde (eds.) (2004: 156–75).

—— MARSDEN, T., and BANKS, J. (2000) 'Quality, Nature and Embeddedness: Some Theoretical Considerations in the Context of the Food Sector', *Economic Geography* 76(2): 107–25.

NESTLE, M. (2002) *Food Politics: How the Food Industry Influenced Nutrition and Health* (Berkeley, Calif.: University of California Press).

Network of GMO-Free Regions (2003) 'Co-existence of Genetically Modified Crops with Traditional and Organic Farming' (Brussels: Assemblée des Régions d'Europe).

NFFC (National Family Farm Coalition) (2002) 'Family Farmers Express Strong Opposition to 2002 Farm Bill', Press Release (Washington, DC: National Family Farm Coalition).

NFU (National Farmers' Union) (2004) 'Review of EU Food Labelling Legislation', Memorandum to Food Standards Agency, 19 August (London: National Farmers Union).

NYGARD, B., and STORSTAD, O. (1998) 'De-globalisation of Food Markets? Consumer Perceptions of Safe Food: The Case of Norway', *Sociologia Ruralis* 38(1): 35–53.

Organic Centre Wales (2004) *Organic Farming in Wales 2003–2004* (Aberystwyth: Organic Centre Wales).

Oxfam (2002) *Rigged Rules and Double Standards: Trade, Globalisation and the Fight Against Poverty* (Oxford: Oxfam International).

—— (2003) 'EU CAP Reforms a Disaster for the Poor', Press Release, 27 June (Oxford: Oxfam).

—— (2004) 'Dumping on the World: How EU Sugar Policies Hurt Poor Countries', *Briefing Paper* 61 (Oxford: Oxfam).

Oxfam (2005) 'Kicking Down the Door: How Upcoming WTO Talks Threaten Farmers in Poor Countries', *Briefing Paper* 72 (Oxford: Oxfam).

PAARLBERG, R. L. (1989) 'The Political Economy of American Agricultural Policy: Three Approaches', *American Journal of Agricultural Economics* 71(5): 1157–64.

PAGE, B. (1996) 'Across the Great Divide: Agriculture and Industrial Geography', *Economic Geography* 72(4), 376–97.

—— (1997) 'Restructuring Pork Production, Remaking Rural Iowa', in Goodman and Watts (eds.) (1997: 133–57).

PARROTT N., WILSON, N., and MURDOCH, J. (2002) 'Spatializing Quality: Regional Protection and the Alternative Geography of Food', *European Urban and Regional Studies* 9(3): 241–61.

PEREZ, J. (2004) 'Community Supported Agriculture on the Central Coast: the CSA Grower Experience', Research Brief 4 (Santa Cruz, Calif.: The Center for Agroecology and Sustainable Food Systems, University of California, Santa Cruz).

PHILLIPS OF WORTH MATRAVERS, LORD, BRIDGEMAN, J., and FERGUSON-SMITH, M. (2000) *The BSE Inquiry Report*, 16 vols. (London: Stationery Office).

PICCHI, A. (2002) 'Rural Development, Institutions, and Policy', in J. van der Ploeg, A. Long, and J. Banks (eds.), *Living Country sides: Rural Development Processes in Europe: The State of the Art* (Doetinchem: Elsevier).

PODBURY, T. (2000) 'US and EU Agricultural Support: Who Does Benefit?', *ABARE Current Issues* 2000(2).

POTTER, C. (1998) *Against the Grain: Agri-environmental Reform in the US and the EU* (Wallingford: CABI).

—— and BURNEY, J. (2002) 'Agricultural Multi-functionality in the WTO: Legitimate Non-trade Concern or Disguised Protectionism?', *Journal of Rural Studies* 18(1): 35–47.

PRATT, J. (1994) *The Rationality of Rural Life: Economic and Cultural Change in Tuscany* (Chur: Harwood Academic).

PRETTY, J. (1998) *The Living Land: Agriculture, Food and Community Regeneration in Rural Europe* (London: Earthscan).

—— (2002) *Agri-Culture: Reconnecting People, Land and Nature* (London: Earthscan).

—— BALL, A. S., LANG, T., and MORISON, J. I. L. (2005) 'Farm Costs and Food Miles: An Assessment of the Full Cost of the UK Weekly Food Basket', *Food Policy* 30(1): 1–19.

PURVIS, A. (2004) 'Loaded: Why Supermarkets Are Getting Richer and Richer', *Observer Food Monthly*, January: 32–8.

RAYNOLDS, L. (1997) 'Restructuring National Agriculture, Agro-food Trade, and Agrarian Livelihood in the Caribbean', in Goodman and Watts (eds.) (1997: 119–32).

—— (2003) 'Forging New Local/Global Links through Fair Trade Agro-food Networks', in R. Almås and G. Lawrence (eds.) *Globalization, Localization and Sustainable Livelihoods* (Basingstoke: Ashgate).

—— MYRHE, D., MCMICHAEL, P., CARRO-FIGUEROA, V., and BUTTEL, F. H. (1993) 'The "New" Internationalization of Agriculture: A Reformulation', *World Development* 21(7): 1101–21.

REARDON, T., TIMMER, C. P., BARRETT, C. B., and BERDEGUÉ, J. (2003) 'The Rise of Supermarkets in Africa, Asia, and Latin America', *American Journal of Agricultural Economics* 85(5) 1140–6.

REEVES, M., et al. (1999) *Fields of Poison: California Farm Workers and Pesticides* (San Francisco: PAN North America, California Legal Assistance Foundation, United Farm Workers of America, and Californians for Pesticide Reform), 21–5.

Regione Toscana (2000) *Rapporto 2000: Rapporto sullo stato dell'ambiente in Toscana* (Florence: Agenzia Regionale per la Protezione Ambientale della Toscana).

RENTING, H., MARSDEN, T. K., and BANKS, J. (2003) 'Understanding Alternative Food Networks: Exploring the Role of Short Food Supply Chains in Rural Development', *Environment and Planning A* 35(3): 393–411.

REVILL, J. (2004) 'McDonald's Bows to Critics and Slashes Salt Ration', *The Observer*, 7 March.

ROBERTS, D. (2002) 'Examination of Witnesses: Mr. David Roberts', in House of Commons, *The Future of UK Agriculture in a Changing World*, Environment, Food and Rural Affairs Committee, Minutes of Evidence and Appendices, HC 550-II (London: Stationery Office).

ROBERTS, R. (2001) 'Examination of Witnesses: Mr. Rees Roberts', in House of Commons. *Farming and Food Policy in Wales*, Welsh Affairs Committee, Minutes of Evidence, HC 427-II (London: Stationery Offce).

ROCH, C. H., SCHOLZ, J. T., and McGRAW, K. M. (2000) 'Social Networks and Citizen Response to Legal Change', *American Journal of Political Science* 44(4): 777–91.

RODRIK, D. (2001) 'The Global Governance of Trade: As If Development Really Mattered', Background Paper to the UNDP Project in Trade and Sustainable Human Development (Cambridge, Mass.: John F. Kennedy School of Government, Harvard University).

ROEP, D. (2001) *The Waddengroup Foundation: The Added Value of Quality and Region*. EU Impact Project Working Paper (Wageningen: The Agricultural University).

—— (2002) 'Value of Quality and Region: The Waddengroup Foundation', in J. D. van der Ploeg, A. Long, and J. Banks (eds.), *Living Countrysides: Rural Development Processes in Europe* (Doetinchem: Elsevier), 88–98.

ROWELL, A. (2003) *Don't Worry (It's Safe To Eat): The True Story of GM Food, BSE and Foot and Mouth* (London: Earthscan).

RSPCA (Royal Society for the Prevention of Cruelty to Animals) (2002) 'Memorandum Submitted by the RSPCA', in House of Commons, *The Future of UK Agriculture in a Changing World*, Environment, Food and Rural Affairs Committee, Minutes of Evidence and Appendices, HC 550-II (London: Stationery Office).

SAGE, C. (2003) 'Social Embeddedness and Relations of Regard: Alternative "Good Food" Networks in South-West Ireland', *Journal of Rural Studies* 19(1): 47–60.

SALAIS, R., and STORPER, M. (1992) 'The Four Worlds of Contemporary Industry', *Cambridge Journal of Economics* 16(2): 169–93.

SANTAGATA, W. (2002) 'Cultural Districts, Property Rights and Sustainable Economic Growth', *International Journal of Urban and Regional Research* 26(1): 9–23.

Sassatelli, R. (2004) 'The Political Morality of Food: Discourses, Contestation and Alternative Consumption', in Harvey, McMeekin, and Warde (eds.) (2004: 176–91).

Sayer, A. (1997) 'The Dialectic of Culture and Economy', in R. Lee and J. Wills (eds.), *Geographies of Economies* (London: Arnold).

—— (2000) 'Moral Economy and Political Economy', *Studies in Political Economy* 62: 79–104.

Schilling, E. (1995) 'Organic Agriculture Grows Up', *California Journal*, May: 21–5.

Sevilla Guzmán, E., and Woodgate, G. (1999) 'Sustainable Rural Development: From Industrial Agriculture to Agro-ecology', in M. R. Redclift and G. Woodgate (eds.), *The International Handbook of Environmental Sociology* (Cheltenham: Edward Elgar).

Shell (2002) *The Quiet Revolution: Ten Years Since Agenda 21* (Birmingham: Shell).

Shell, E. R. (2002) *The Hungry Gene: The Science of Fat and the Future of Thin* (New York: Atlantic Monthly Press).

Smith, E. (2002) 'Ecological Modernisation and Organic Farming in the UK: Does it Pay to be Green?' Unpublished Ph.D. Dissertation, Cardiff University.

—— and Marsden, T. (2004) 'Exploring the "Limits to Growth" in UK Organics: Beyond the Statistical Image', *Journal of Rural Studies* 20(3): 345–57.

—— —— Flynn, A., and Percival, A. (2004) 'Regulating Food Risks: Rebuilding Confidence in Europe's Food?', *Environment and Planning C: Government and Policy* 22(4): 543–67.

Sonnino, R. (2003) 'For a "Piece of Bread"? Interpreting Sustainable Development through Agritourism in Southern Tuscany, Italy', Ph.D. thesis, University of Kansas, Lawrence.

Spencer, C. (2002) *British Food: An Extraordinary Thousand Years of History* (London: Grub Street).

Stoll, S. (1998) *The Fruits of Natural Advantage: Making the Industrial Countryside in California* (Berkeley, Calif.: University of California Press).

Storper, M. (1997). *The Regional World: Territorial Development in a Global Economy* (London: Guilford).

—— and Salais, R. (1997) *The Worlds of Production: The Action Frameworks of the Economy* (Cambridge, Mass.: Harvard University Press).

Straete, E. P., and Marsden, T. (forthcoming) 'Exploring the Dimensions of Designed Qualities in Food'. *Economic Geography*.

Stuiver, M., and Marsden, T. K. (forthcoming) 'The Promise of Retro-innovation for Rural Development: Theorising beyond the Modernisation Paradigm'.

Sumner, D. A. (2003) 'Implications of the US Farm Bill of 2002 for Agricultural Trade and Trade Negotiations', *Australian Journal of Agricultural and Resource Economics* 46(1): 99–122

—— and Brunke, H. (2004) 'Commodity Policy and California Agriculture', in J. Siebert (ed.) *California Agriculture: Dimensions and Issues* (Oakland, Calif.: ANR Communication Services), 157–80.

—— and Wolf, C. A. (1996) 'Quotas without Supply Control: Effects of Dairy Quota Policy in California', *American Journal of Agricultural Economics* 78(2): 354–66.

SWYNGEDOUW, E. (2000) 'Authoritarian Governance, Power and the Politics of Rescaling', *Environment and Planning D: Society and Space* 18(1): 63–76.

SYLVANDER, B., ALBISU, L. M., ALLAIRE, G., ARFINI, F., BARJOLLE, D., BELLETTI, G., CASABIANCA, F., LASSAUT, B., MARESCOTTI, A., THÉVENOD-MOTTET, E., and TREGEAR, A. (2004) 'Final Report: Synthesis and Recommendations', DOLPHINS WP7 (Le Mans: INRA-UREQUA).

TEGTMEIER, E., and DUFFY, M. (2004) 'External Costs of Agricultural Production in the United States', *International Journal of Agricultural Sustainability* 2(1): 1–20.

THEVENOT, L., MOODY, M., and LAFAYE, C. (2000) 'Forms of Valuing Nature: Arguments and Modes of Justification in French and American Environmental Disputes', in M. Lamont and L. Thevenot (eds.), *Rethinking Comparative Cultural Sociology: Repertoires of Evaluation in France and the United States* (Cambridge: Cambridge University Press), 229–72.

THORNTON, G. (1996) 'High Yielding Broiler Industry: The Big Trade-off', *Broiler Industry*, August.

TREGEAR, A. (2003) 'From Stilton to Vimto: Using Food History to Re-think Typical Products in Rural Development', *Sociologia Ruralis* 43(2): 91–107.

UBS Warburg (2002) *Absolute Risk of Obesity* (London: UBS Warburg).

USDA (United States Department of Agriculture) (2001) *Food and Agricultural Policy: Taking Stock for the New Century* (Washington, DC: USDA).

VAN DER GRIJP, N., MARSDEN, T., and CAVALCANTI, J. S. B. (forthcoming) 'European Retailers as Agents of Change towards Sustainability: The Case of Fruit Production in Brazil', *Environmental Sciences*.

VAN DER PLOEG, J. D. (2003) *The Virtual Farmer: Past, Present and Future of the Dutch Peasantry* (Assen: Royal Van Gorcum).

—— (in press) 'Agricultural Production in Crisis', in P. Cloke, T. Marsden, and P. Mooney (eds.) *Handbook of Rural Studies* (London: Sage).

—— LONG, A., and BANKS, J. (eds.) (2002) *Living Countrysides: Rural Development Processes in Europe* (Doetinchem: Elsevier).

VICTOR, D. G., and RUNGE, C. F. (2003) 'A Trade Battle That Will Cost America Dear', *Financial Times*, 15 May.

WAITROSE (2004) *Corporate Social Responsibility Report 2004* (Bracknell: Waitrose).

WALKER, R. A. (1999) 'Putting Capital in Its Place: Globalization and the Prospects for Labor', *Geoforum* 30(3): 263–84.

—— (2001) 'California's Golden Road to Riches: Natural Resources and Regional Capitalism, 1848–1940', *Annals of the Association of American Geographers*, 91(1): 167–99.

—— (2004) *The Conquest of Bread: 150 Years of Agribusiness in California* (New York: New Press).

WARD, N. (2003) 'Environmental Policy Integration and Agriculture in Europe', paper presented at international conference on Environmental Policy Integration and Sustainable Development, ANU, Canberra, November.

—— and ALMÅS, R. (1997) 'Explaining Change in the International Agro-food System', *Review of Intrnational Political Economy* 4(4): 611–29.

—— and FALCONER, K. (1999) 'Greening the CAP through Modulation, Opportunities and Constraints', *Ecos* 20(2): 43–50.

WARD, N. and LOWE, P. (2002) 'Devolution and the Governance of Rural Affairs in the UK', in J. Adams and P. Robinson (eds.) *Devolution in Practice* (London: IPPR), 117–39.

WATKINS, K. (2003) 'Reducing Poverty Starts With Fairer Farm Trade', *Financial Times*, 2 June.

WATTS, D. C. H., ILBERY, B., and MAYE, D. (2005) 'Making Reconnections in Agro-food Geography: Alternative Systems of Food Provision', *Progress in Human Geography* 29(1): 22–40.

WATTS, G., and KENNETT, C. (1995) 'The Broiler Industry', *The Poultry Tribune* 9/95: 6–18.

WAG (Welsh Assembly Government) (2000) *Learning to Live Differently* (Cardiff: WAG).

WAG (Welsh Assembly Government) (2001) *Farming for the Future: A New Direction for Farming in Wales* (Cardiff: Welsh Assembly Government).

WDA (Welsh Development Agency) (2004) *Agri-Food Strategy* (Cardiff: WDA).

WELLS, M. J. (1998) *Strawberry Fields: Politics, Class, and Work in Californian Agriculture* (Ithaca, NY: Cornell University Press).

—— and VILLAREJO, D. (2004) 'State Structures and Social Movement Strategies: The Shaping of Farm Labor Protections in California', *Politics and Society* 32(3): 291–326.

WHATMORE, S. (1994) 'Global Agro-food Complexes and the Refashioning of Rural Europe', in A. Amin and N. Thrift (eds.) *Globalization, Institutions, and Regional Development* (Oxford: Oxford University Press).

—— and THORNE, L. (1997) 'Nourishing Networks: Alternative Geographies of Food', in Goodman and Watts (eds.) (1997: 176–91).

—— STASSART, P., and RENTING, H. (2003) 'What's Alternative about Alternative Food Networks?', *Environment and Planning A* 35(3): 389–91.

WHIRTY, R. (2000) 'Genetic Engineering of Strawberries Won't Be Easy Task', *The Fruit Growers News* 12.

WILKINSON, J. (1997a) 'A New Paradigm for Economic Analysis?', *Economy and Society* 26(3): 305–39.

—— (1997b) 'Regional Integration and the Family Farm in the Mercosul Countries: New Theoretical Approaches as Supports for Alternative Strategies', in Goodman and Watts (eds.) (1997: 35–64).

WINTER, M. (2003) 'Embeddedness, the New Food Economy and Defensive Localism', *Journal of Rural Studies* 19(1): 23–32.

World Health Organization (1998) *Obesity: Preventing and Managing the Global Epidemic* (Geneva: WHO).

WRIGLEY, N. (2002) 'Transforming the Corporate Landscape of US Food Retailing: Market Power, Financial Re-engineering and Regulation', *Tijdschrift voor economische en sociale geografie* 93(1): 62–82.

—— and LOWE, M. (2002) *Reading Retail: A Geographical Perspective on Retailing and Consumption Spaces* (London: Arnold).

WRONG, M. (2000) 'Consumers "suspicious and fearful" of food they eat', *Financial Times*, 23 May.

WTO (World Trade Organization) (2001) *Doha Ministerial Declaration* (Geneva: WTO).

—— (2002) *The Doha Declaration Explained* (Geneva: WTO).

YOUNG, A. (2003) 'Political Transfer and Trading Up? Transatlantic Trade in GM Food and US Politics', *World Politics* 55(4): 457–84.

YOUNG, W. (2004) *Sold Out! The True Cost of Supermarket Shopping* (London: Vision).

INDEX